MathWise Decimals

With Answer Key

Peter Wise

MATH TEACHER, MONUMENT, COLORADO

CONTRIBUTORS

David Wise

Katherine Wise

Cover Design by Kris Budi

Dedicated to my wife, Allison, and to my two children, David and Katie.

Special thanks to Shi Hayes for her helpful suggestions and for proofreading the introduction. Thanks also to my principal, Aileen Finnegan, for her leadership, support, and encouragement.

MathWise Decimals with Answer Key

MathWise Curriculum Press

First printing 2014

MathWise Decimals

TABLE OF CONTENTS

HOW TO USE THIS BOOK

While students can use this series profitably when working alone, my experience has indicated that they will make greater progress if guided by a parent, tutor, or helper. This is particularly true for younger students. This person need not be a math teacher at all—just a reader.

If a student or parent is unclear about a solution or procedure, he or she is encouraged to check with the answer key (available separately).

Web Site
For questions, comments, or suggestions, please visit www.mathwisebooks.com.

POINTS OF EMPHASIS

Too many students learn math as if they were learning a dead language. To them math consists of memorizing a bunch of rules and formulas—definitely a wrong approach! Rather, it is important for people to know **how and why math works the way it does.** Students need to be trained to think mathematically from preschool through college, in every grade level. A formulaic understanding of math is both harder to learn and easier to forget.

Tips and tricks help as memory aids and have a legitimate role in acquiring and retaining information. However it is even more important that students understand the reasoning behind rules and formulas. **The MathWise series incorporates both tips/tricks as well as reasoning behind math formulas and procedures.**

FOR WHICH GRADE LEVEL(S) ARE THESE BOOKS INTENDED?

This series is based on skill sets, not grade or age. The workbooks are intentionally created to be suitable for a wide range of grades. The focus is upon math topics, not a grade-level curriculum. If students need extra support in a given topic area, these books are designed to enrich any curriculum they may be using. If a student in middle or high school is rusty on a skill set normally presented in earlier grades, this series will address this area of need, without stigmatizing the student. On the other hand, if a student in lower grades has a parent or teacher who is willing to walk the student through the explanations and exercises in the book, he or she will also profit from the study. Front-loading key math concepts will make future math classes that much easier.

The source materials for these books were used in several different grade-level classes, and even multi-level before- and after-school math clubs. The intention is that solid math is taught without distracting students with the barriers of grade distinctions.

SPECIAL FEATURES OF THIS BOOK

This series of books is designed to be unique and to catch kids' attention in special ways:

Tips and Tricks
Over the years, I have assembled a wide assortment of memory aids—my tips and tricks. Students have found these to be helpful and memorable, but they have also found that these pointers add interest and excitement to their math studies.

Speech Bubbles with Teacher Insights
Speech bubbles are used to provide guidance, point out insights, or give helpful hints as students are solving math problems. Students learn best by doing, and the instruction given in the speech bubbles is designed to (1) sharpen students' powers of observation, (2) increase number sense, and (3) instruct in digestible chunks.

Higher-Order Mathematical Thinking Skills
Rather than depending solely on rote memorization, this series endeavors to explain to students how and why the math does what it does all along the way. This increases retention of concepts and fosters a deeper understanding of mathematics

Simplicity of Instruction
Concepts are explained clearly and simply. Every page (excluding review pages or concept quizzes) has a specific focus. Most pages have generous amounts of white space to keep students focused. Movement is from the simple to the increasingly complex.

Step-By-Step Procedures
Students learn best when given explicit, step-by-step instruction. When several steps are involved, they are numbered. This makes learning much more logical and memorable.

Depth and Complexity
Throughout the book there are challenge problems to stretch students' thinking. At your discretion, you can guide students through the more challenging problems (recommended) or, alternatively, you can skip these harder problems.

Informal Terms
This book often employs informal language like "top number" or "bottom number" to keep things simple and focused. Standard mathematical terminology, such as numerator and denominator, is used after the concepts are presented.

Logical-Sequential Instruction
Math problems are presented in a logical sequence so that previous problems contribute to students' abilities to solve future problems. The order in which you present math problems is critical to promoting number sense.

TEACHING DECIMALS

Importance of Fraction-Decimal-Percent Conversions
I often tell my students that they can think of numbers as wearing three different costumes at this point: fraction costumes, decimal costumes, and percent costumes. The exact same "actor" (number) can slip into different costumes, but he remains the same person. Students need to be able to help numbers to change outfits; that is, converting fractions to decimals, decimals to percents, and so on. Proficiency in these conversions is an integral part of all math.

Place Value and "The Rainbow Chart"
Decimals play an important role in understanding place value—a critically important and foundational skill for understanding our number system and developing number sense. Decimals are best understood in the context of our base-10 system. They are simply numbers divided by powers of tens.

To help students understand place value I developed a model called "The Rainbow Chart" that I have used with success for many years. I start with four boxes to the left of the decimal point and three to the right of it. I circle the ones/units place and tell the students that this place is special, so we are going to circle it. Then I go to the tens place and, after reminding students that any digit in that box is really multiplied by 10, I review the tenths place and draw a connecting ("rainbow") line from the 10s & 10ths place, 100s & 100ths place, and 1000s & 1000ths place.

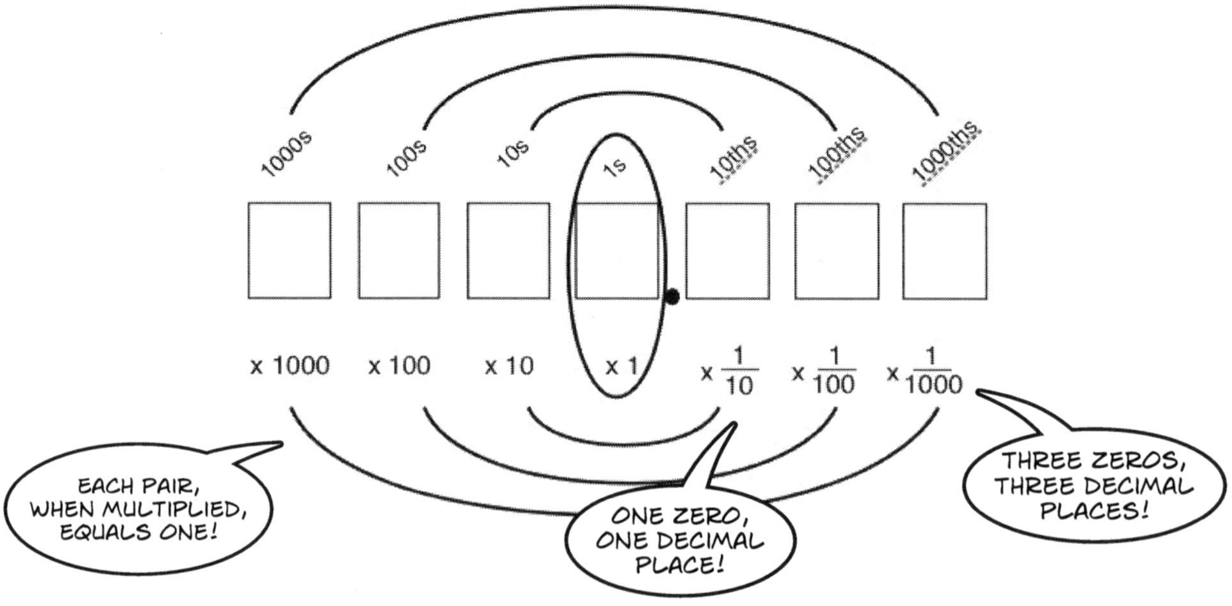

There are several valuable lessons that can be taught below the Rainbow Chart boxes. Under the tenths place, I will often say that any digit here is multiplied by 1/10. I tell them that each Rainbow Pair, (10s & 10ths, 100s & 100ths, etc.) when multiplied together, equals one.

As an alternative, I tell students that under the 10ths place, you could also say that the digit is multiplied by .1, under the 100ths place the digit is multiplied by .01, etc.

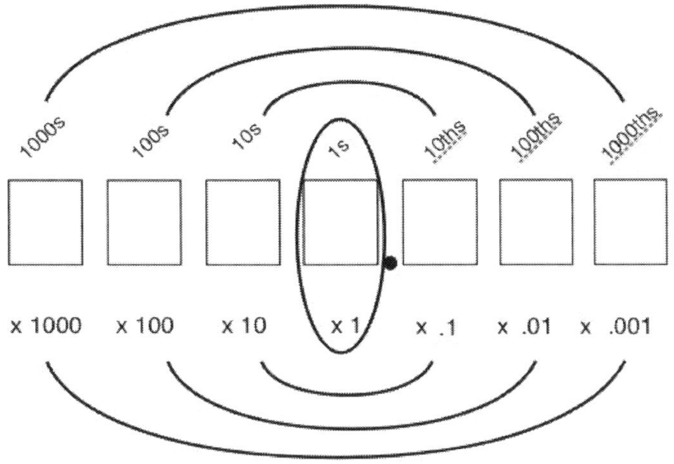

Increasing the level of rigor, I will then put powers of 10 under each Rainbow Chart box.

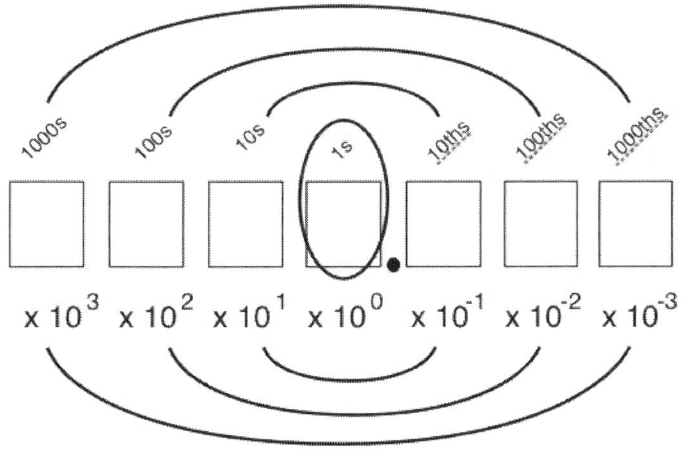

From here, I teach students what I call the "Last Digit Up" trick. In this procedure, decimal numbers are read as regular numbers, then after the last digit, students read the place value above that last digit. (.63 is read first as "sixty-three"), and then students have to see that the last digit—the 3—tells them to point up to the rainbow box name: in this case, hundredths.

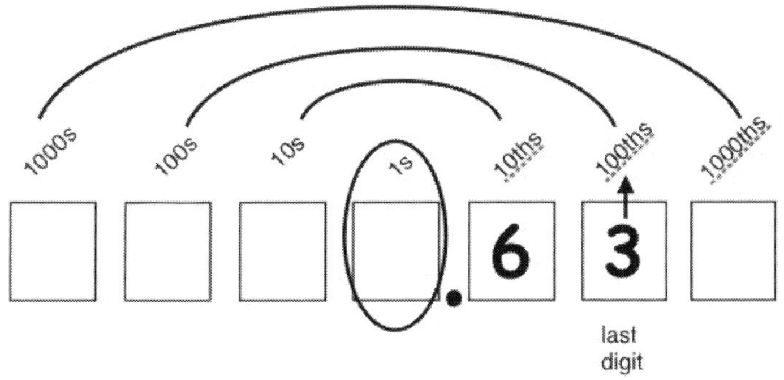

I often go back to the Rainbow Chart and review fundamental principles of our number system with students. At higher levels, I will even introduce students to different base systems (base 8, base 12, etc.). Negative exponents are illustrated well in this model—they do not necessarily mean that their values are negative at all! Rather, negative exponents show division.

After reviewing the Rainbow Chart several times with a fourth grade class, I put 7 in the thousands place and posed the question, "The seven in this place is really multiplied by what power of ten?" Someone replied, "three," and I was ready to move on, but one student raised his hand and asked if he could come to the board and relate his discovery to the class. I consented and was impressed as he presented his idea. He erased the "x 1000" under the 7 that I had written and offered an alternative solution: "That 7 in the 1000s place can also be written as $7 \div 10^{-3}$!" This demonstrated his deep understanding of our number system and place value—particularly at his grade level!

At the back of this book, I have included supplementary pages with a generic sheet for converting decimals, percents, and fractions along with different variations of the rainbow chart. I recommend making copies for your student or class to teach/review the following concepts:

- Place value and rainbow arcs
- Last Digit Up trick for reading, writing, and naming decimals
- Fractions (any number in the tenth's place = x 1/10, etc.)
- Decimals (any number in the tenth's place = x .1, etc.)
- Powers of ten (any number in the tenth's place = x 10^{-1})

Peter Wise
Monument, Colorado
2016

What are Decimals Anyway?

- Decimals are just another way of writing fractions (and percents)

Numbers wear three costumes

A LOT OF MATH INVOLVES LEARNING HOW TO CHANGE COSTUMES!

FRACTIONS	DECIMALS	PERCENTS
$2\frac{1}{10}$	2.1	210%

YOU CAN CALL THEM "FDP"

- The same actor (number) can wear any of these 3 costumes.
- He may look different in these costumes—but he's really the same actor!

- Instead of writing $\frac{1}{10}$, you can just write 1 to the RIGHT of the DECIMAL POINT! (.1)
- Decimals are just ways of writing fractions that have 10, 100, 1000, etc. as the denominator. (You'll learn about this later!)

Change from the FRACTION costume to the DECIMAL costume

1. $\frac{3}{10}$ = . ☐

2. $\frac{9}{100}$ = . 0 ☐

3. $\frac{7}{1000}$ = . 0 0 ☐

4. $\frac{23}{100}$ = . ☐ ☐

Using Decimal Bars

Models can help to visualize decimals

A. Shade the correct number of bars to show .3

B. Shade the correct number of bars to show 1.2

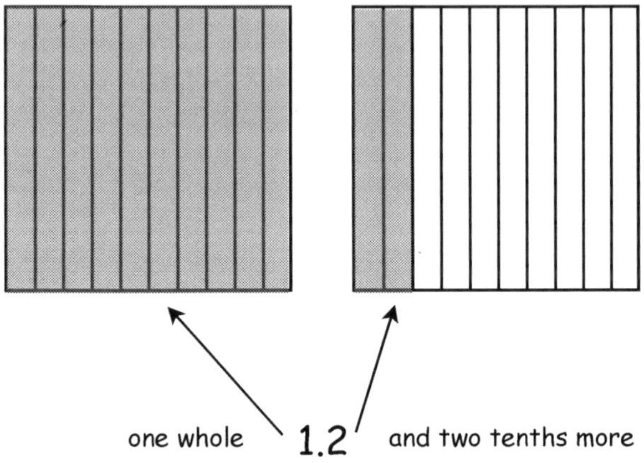

Shade the correct number of bars to show the decimal

1. .4

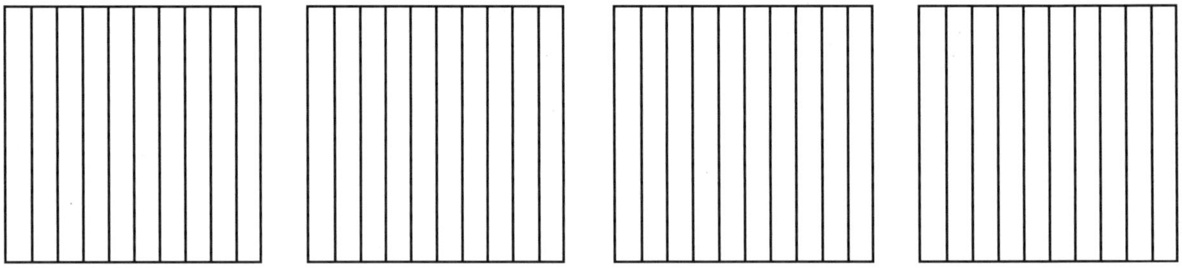

2. 2.8

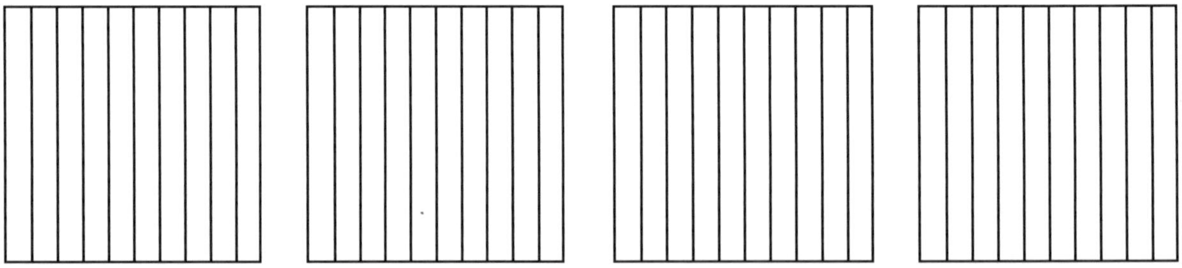

3. 3.7

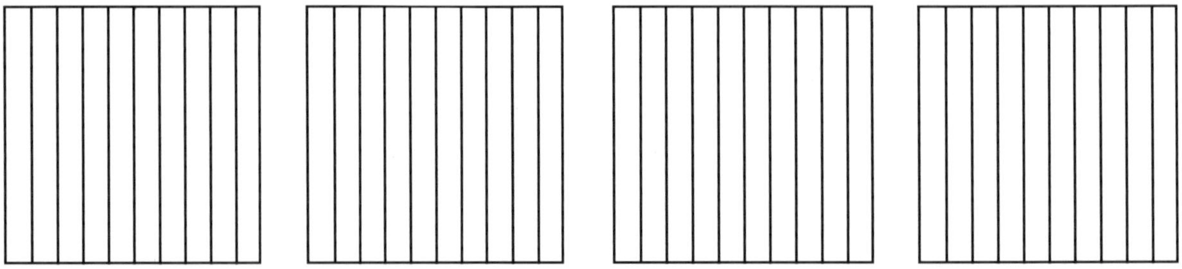

Using Decimal Bars

Write the correct decimal amounts

1. 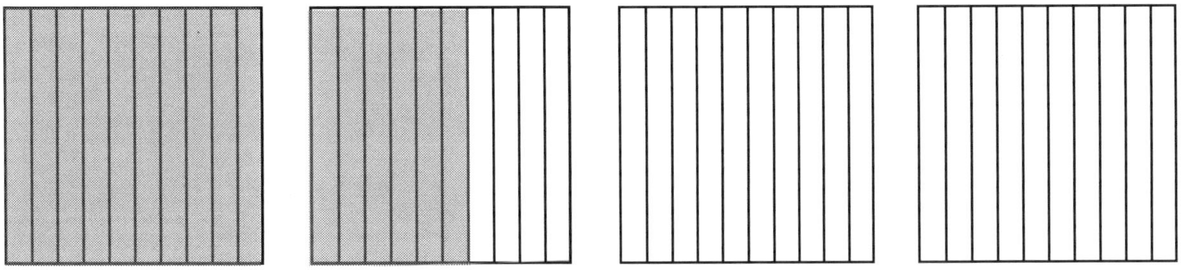

2. 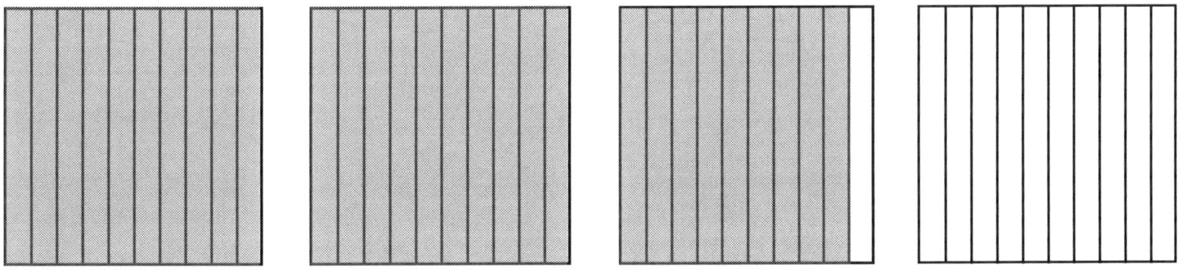

3. 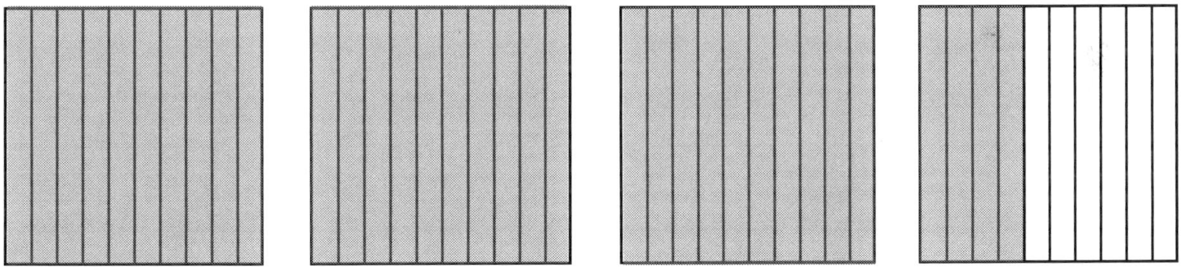

4. 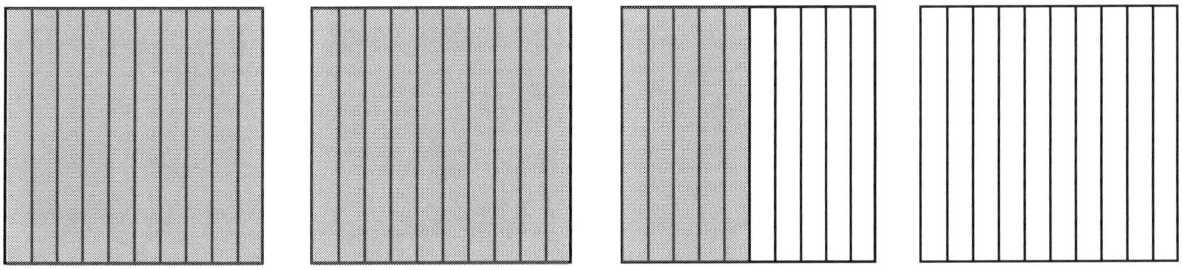

5. 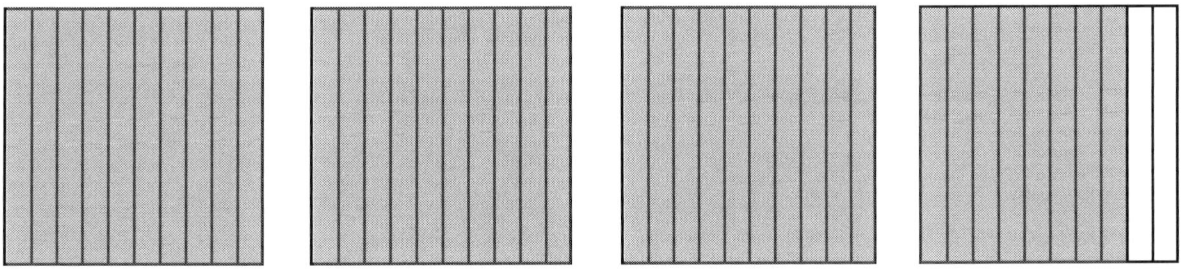

Using Decimal Boxes

A. Shade the correct number of boxes to show .07

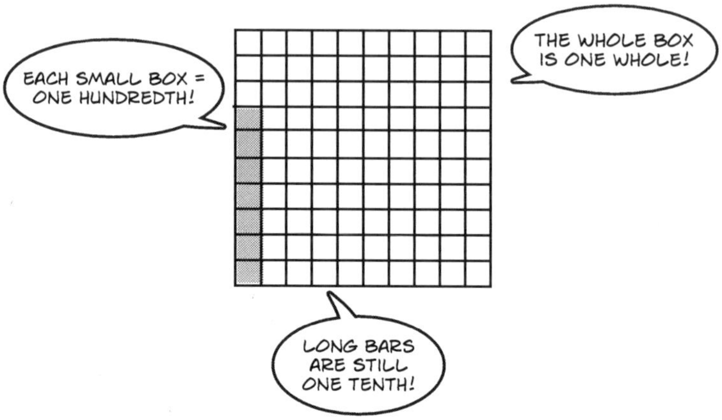

B. Shade the correct number of boxes to show 1.02

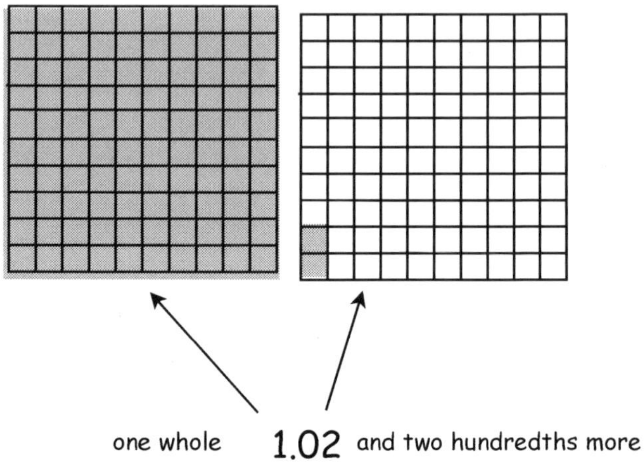

Shade the correct number of boxes to show the decimal

1. .04

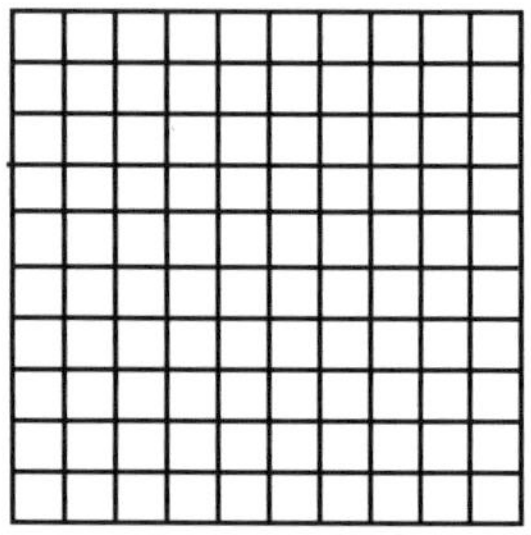

4. .56

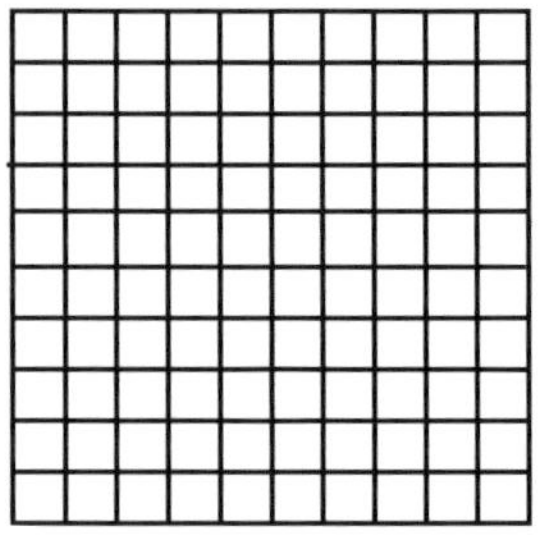

2. .40 (or .4)

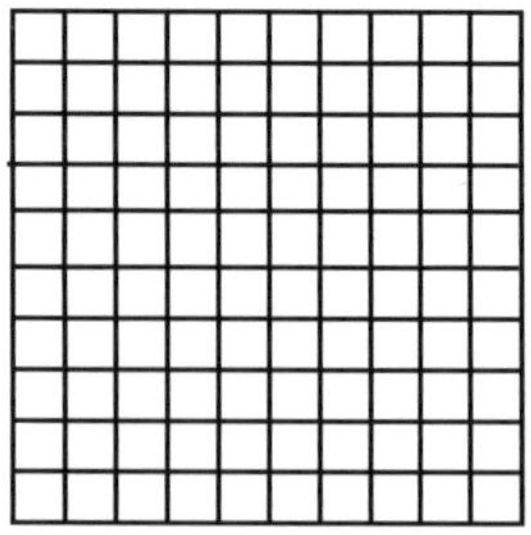

5. .88

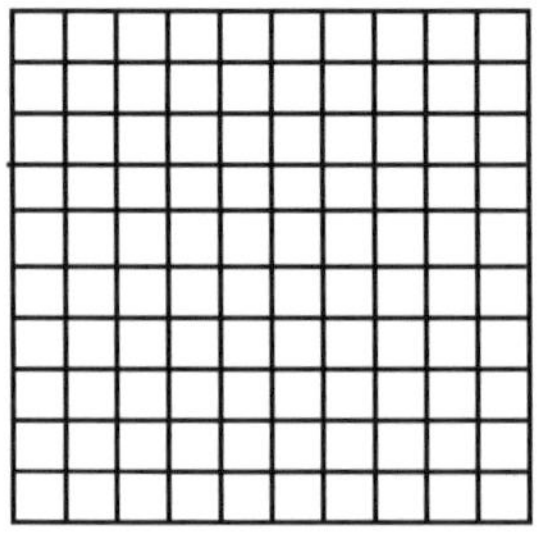

3. 2.75

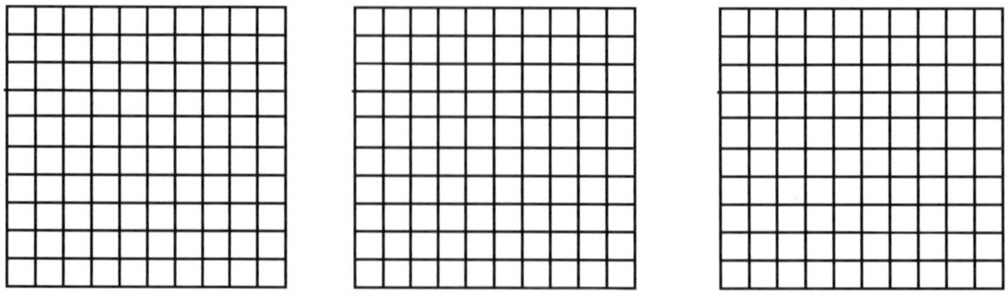

Using Decimal Boxes

Examples

A. .03

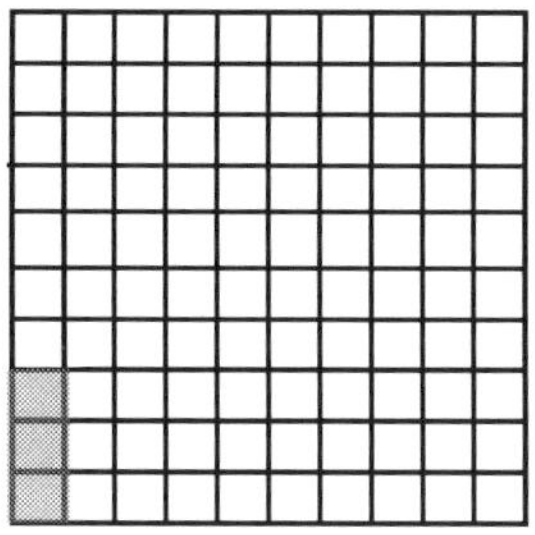

B. .3 (or .30)

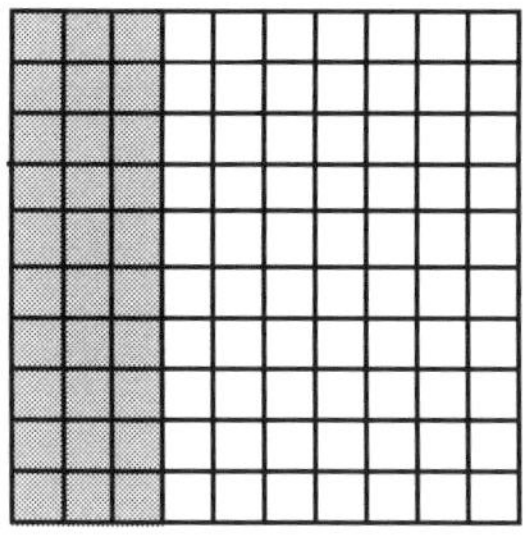

.3 three long bars are the same as...
.30 30 small boxes

Write the correct number for each decimal

1.

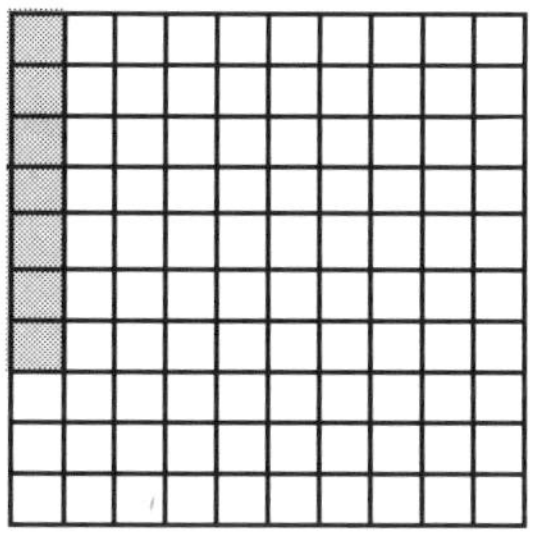

5.

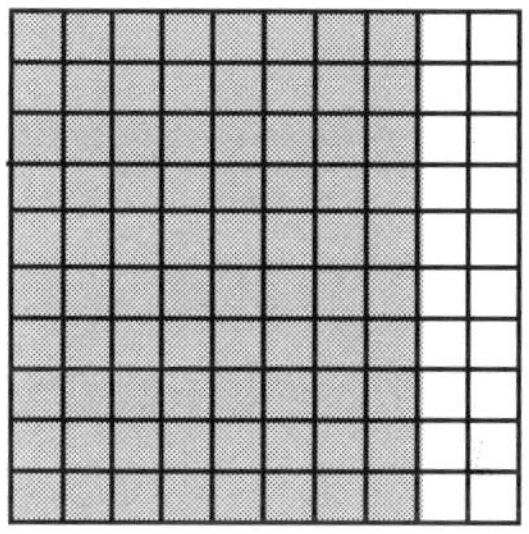

2.

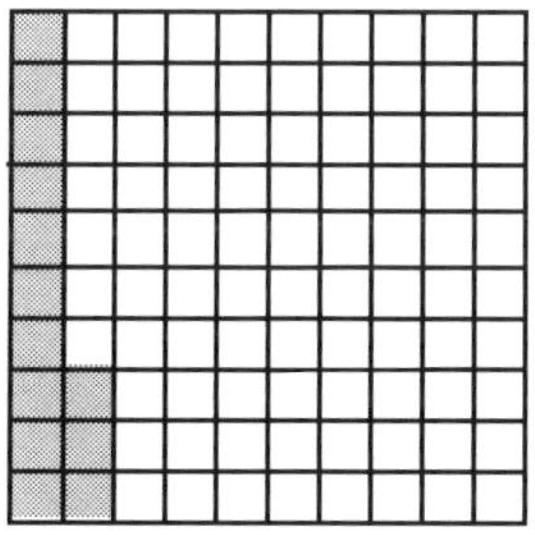

6.

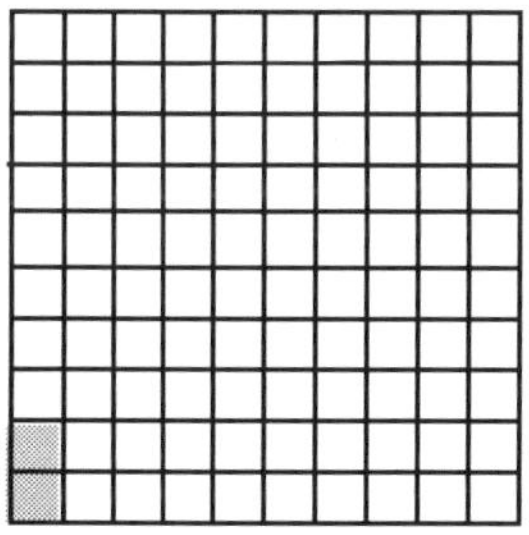

3. or

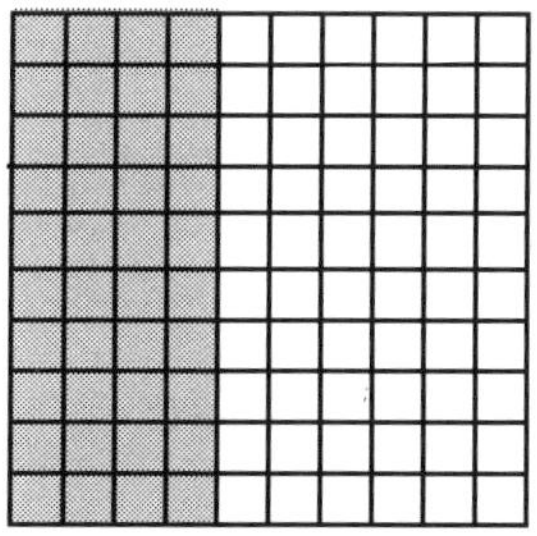

What decimal UNshaded?

7.

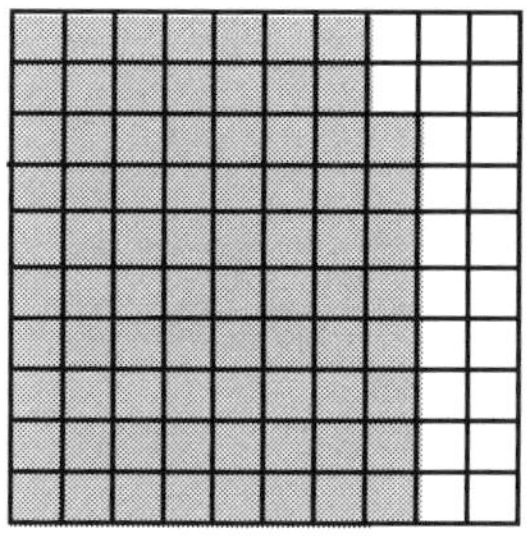

What decimal UNshaded?

4.

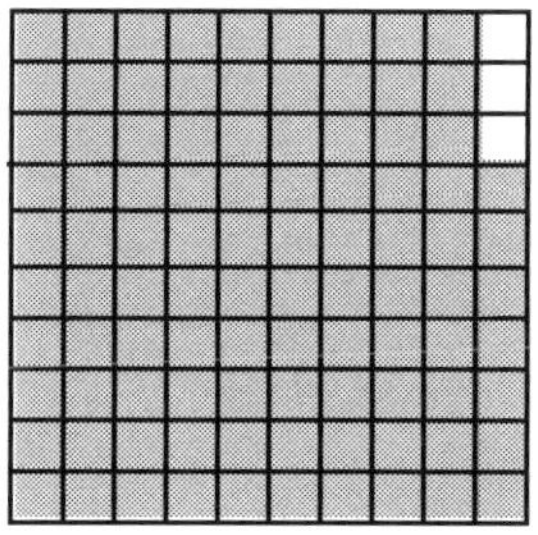

What decimal UNshaded?

8.

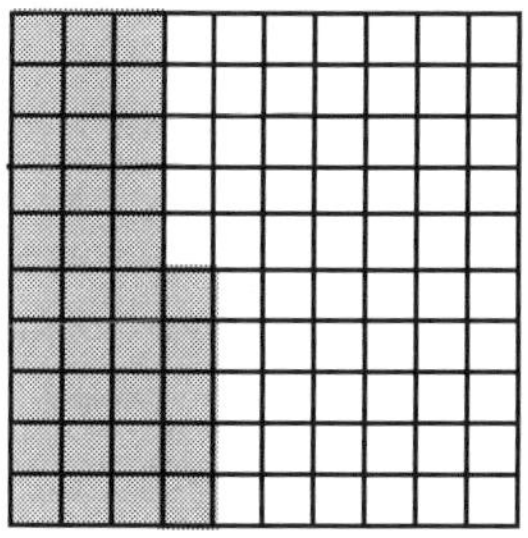

What decimal UNshaded?

Introduction to the Rainbow Chart

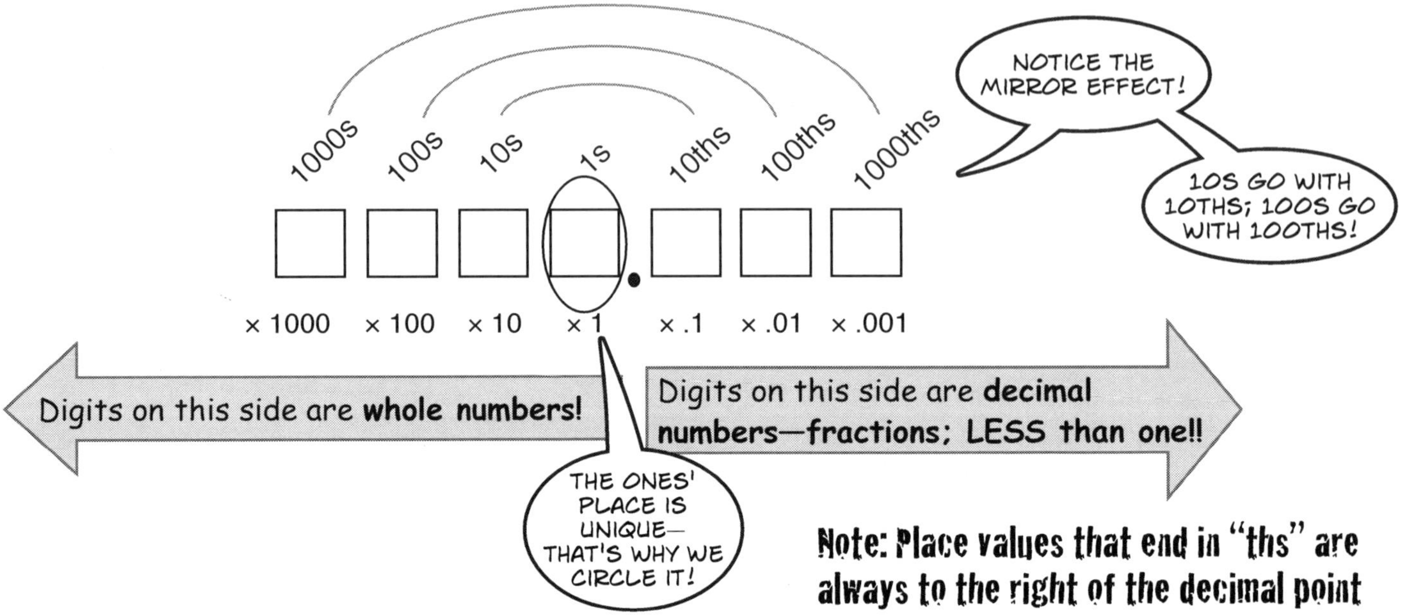

Note: Place values that end in "ths" are always to the right of the decimal point

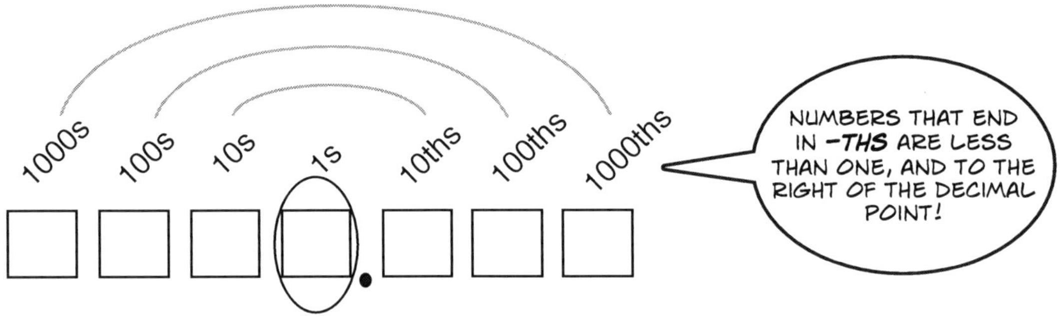

PROBLEM #1: CONNECT THE BOXES WITH RAINBOW LINES. FILL IN THE BOXES AS IN THE EXAMPLES ABOVE.

PROBLEMS 2–3: PUT THE DIGITS IN THE CORRECT PLACES.

1.

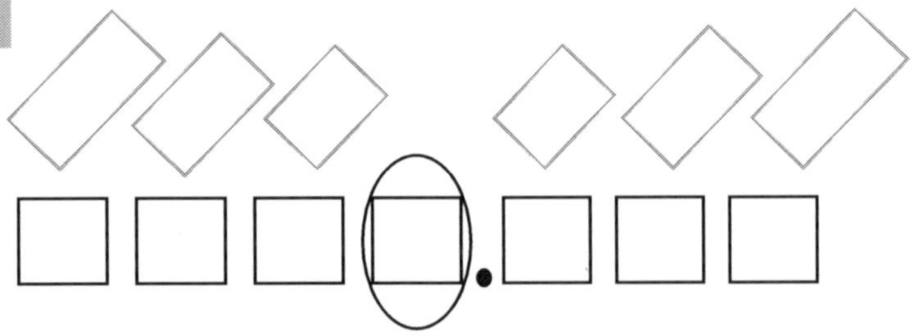

2. Put 3 in the ones place; 7 in the tens place; 8 in the tenths place; 6 in the hundreds place; 9 in the hundredths place

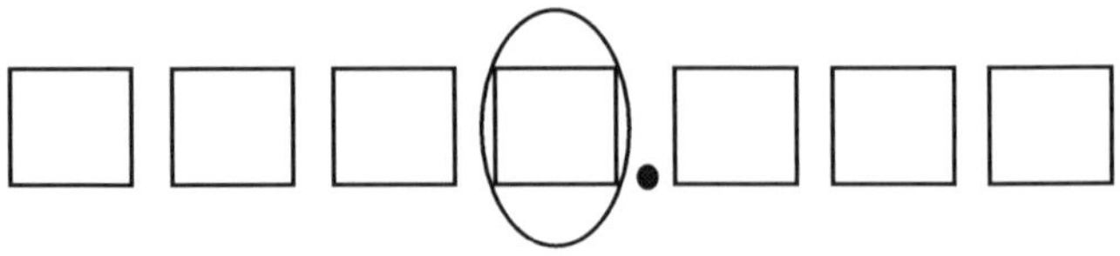

3. Put 4 in the thousands place, 7 in the thousandths place, 5 in the hundreds place, 9 in the hundredths place, 8 in the tens place, 2 in the tenths place and 1 in the ones place →

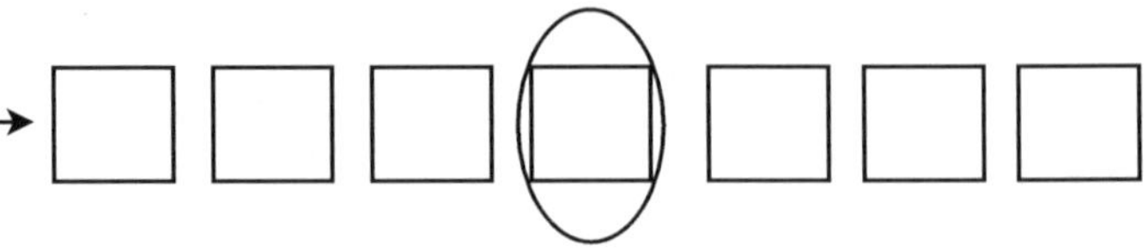

Rainbow Chart

Remember: Place values that end in "ths" are always to the right of the decimal point

1. Label the boxes with the correct place value and draw rainbow lines. Show the multiplication associated with each place value box. Do you see how closely related the names of the rainbow pairs are?

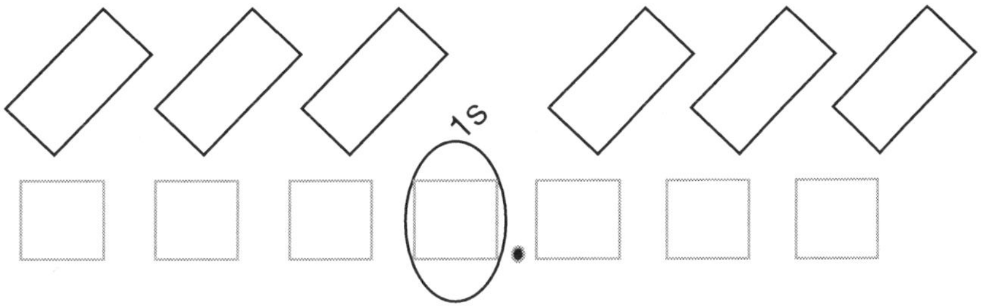

times: ______ times: ______ times: ______ times: ______ times: ______ times: ______

2. Put 6 in the tenths place;
put 2 in the thousandths place.
You can't have a blank in the hundredths place…
so what do you need to put there?

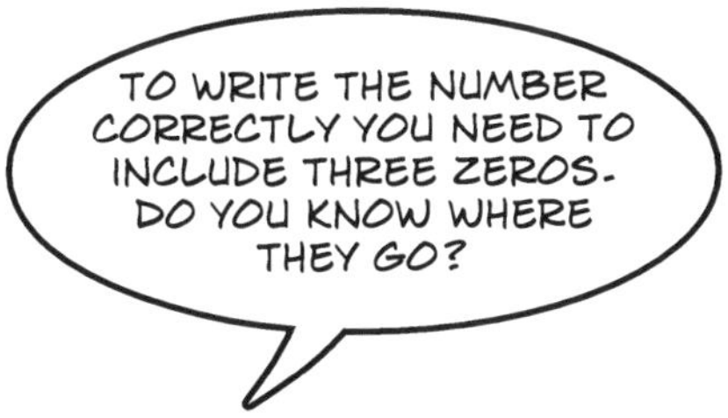

3. Put 7 in the hundreds place;
put 3 in the hundredths place;
put 1 in the thousandths place.

4. Put 6 in the thousandths place;
put 4 in the hundredths place;
put 8 in the tenths place.

NOW PRACTICE WITH PAGES 75–76
(THOSE PAGES ARE REPRODUCIBLE)

Last Digit Up Trick

A. 27 hundredths

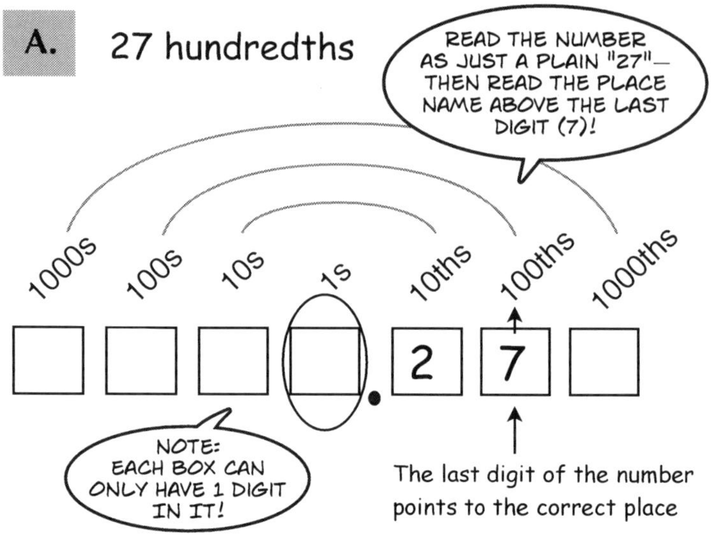

B. 48 thousandths

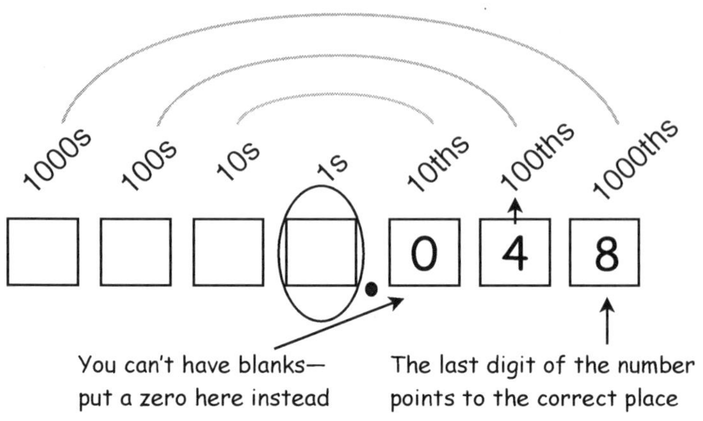

Write the decimals in the correct places

to read the decimals, just read the number, then name the place of the last digit

Write the number 98 so that the last digit (8) points to the hundredths place

1. 98 hundredths

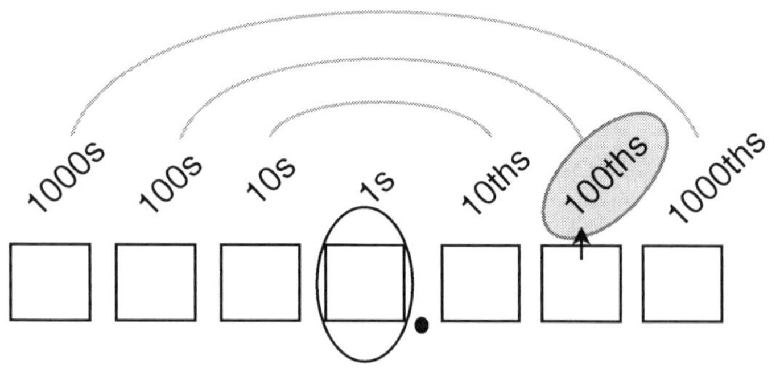

2. 274 thousandths

(Draw rainbow lines below first; then write the number in the correct place)

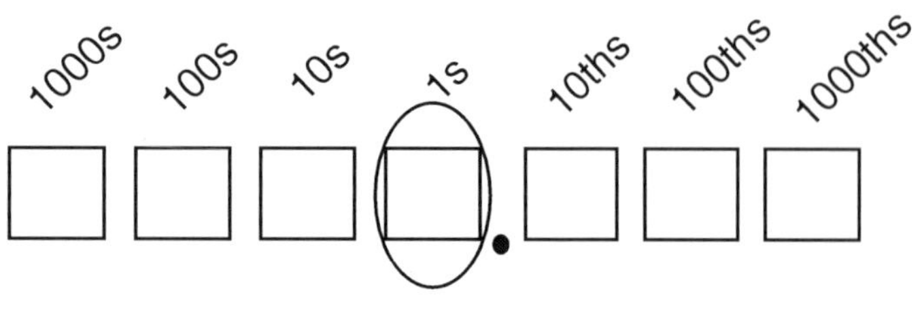

3. 4 thousandths

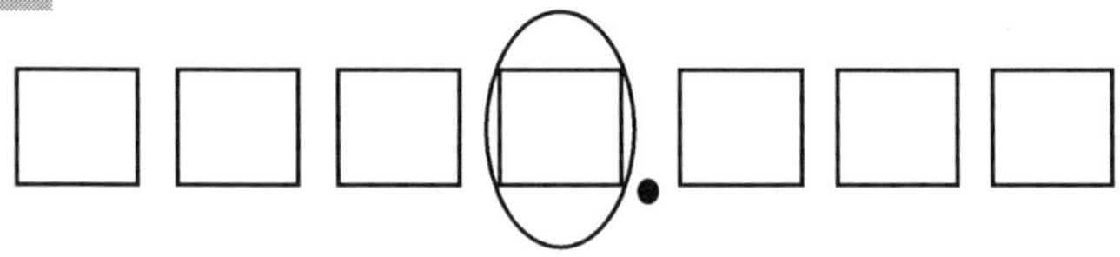

4. 45 thousandths

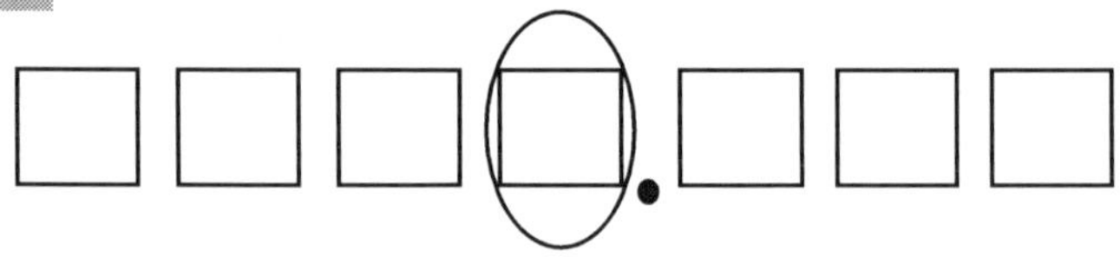

5. 6 hundredths

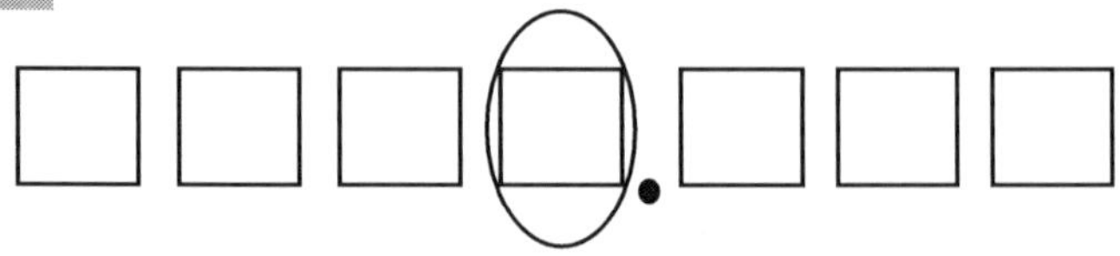

6. 208 thousandths

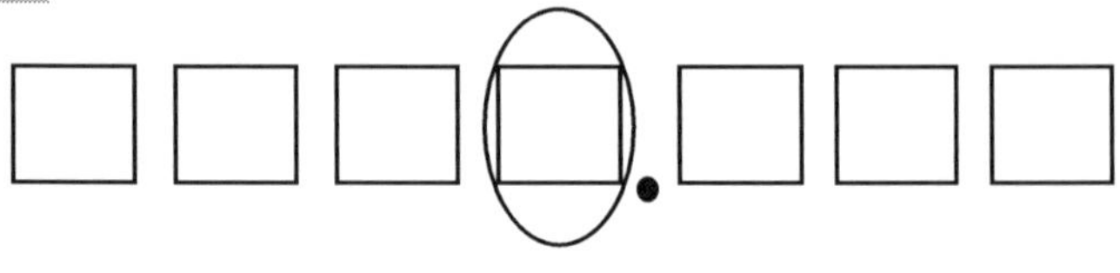

Last Digit Up Trick

A. Circle the place of the last digit

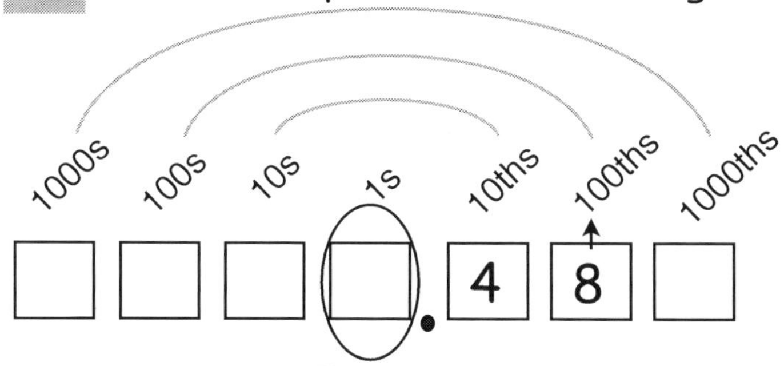

48 tenths hundredths thousandths

THE LAST DIGIT POINTS UP TO THE THOUSANDTHS PLACE!

B. Circle the place of the last digit

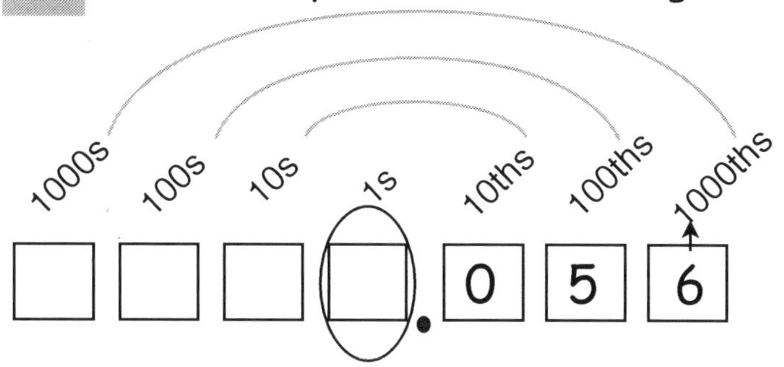

56 tenths hundredths thousandths

Skip the zero (56 is really the same as 056)!

The last digit of the number points to the correct place

Draw rainbow lines, label the places, and circle the correct place for the last digit

1. .23

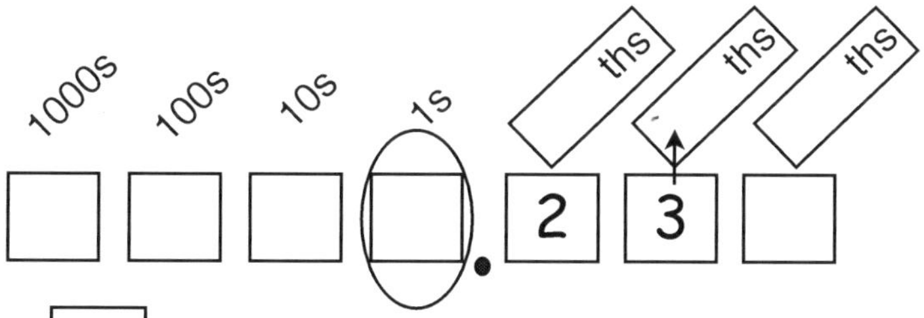

23 tenths hundredths thousandths

2. .865

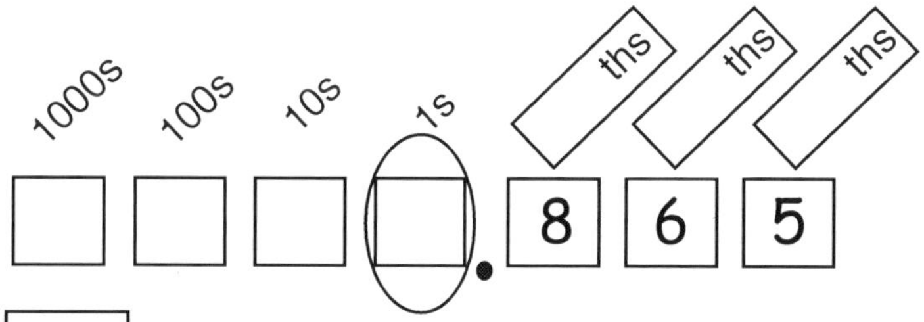

865 tenths hundredths thousandths

3. .078

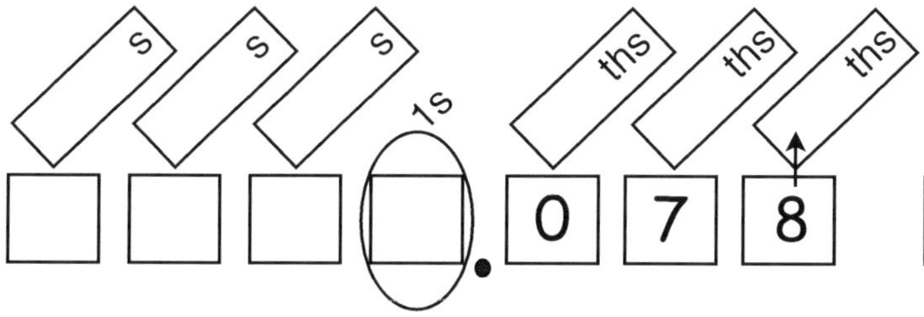

78 tenths hundredths thousandths

4. .309

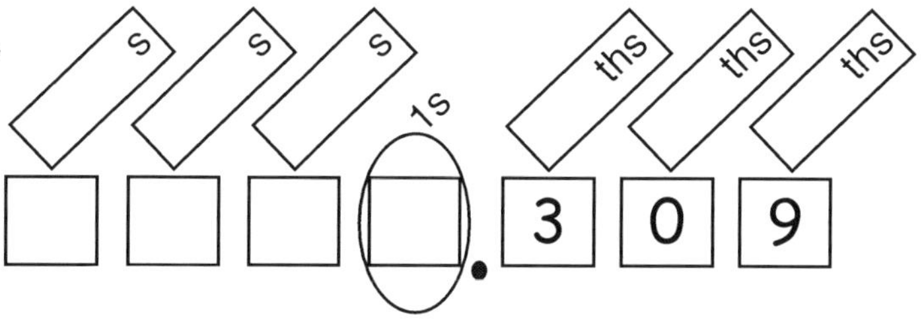

309 tenths hundredths thousandths

Last Digit Up Trick

Write the number and circle the correct place for the LAST DIGIT of each decimal number

A. Circle the place of the last digit

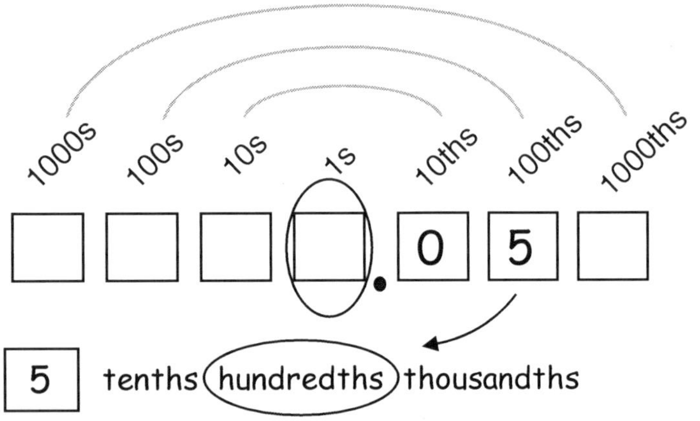

1. Draw rainbow lines, write the number, and circle the place of the last digit

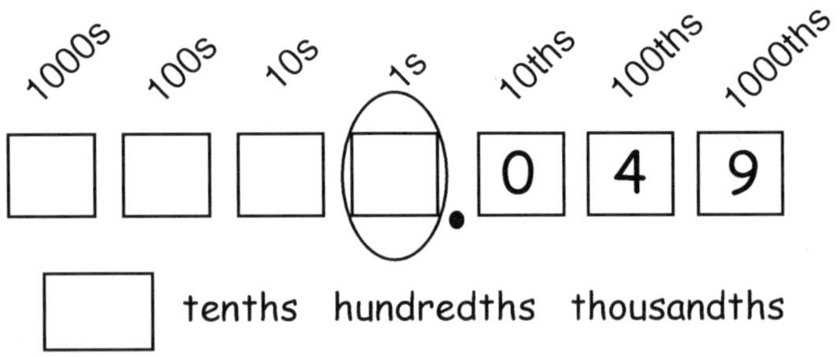

2. (1) Draw rainbow lines and label the places

(2) Write the number and circle the correct place

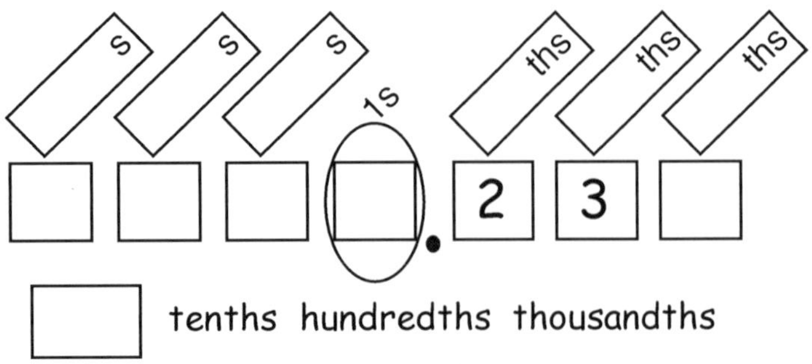

3. (1) Draw rainbow lines and label the places

(2) Write the number and circle the correct place

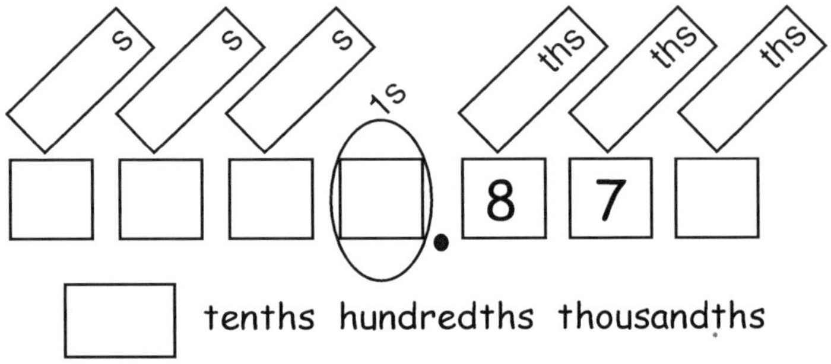

4. (1) Draw rainbow lines and label the places

(2) Write the number and circle the correct place

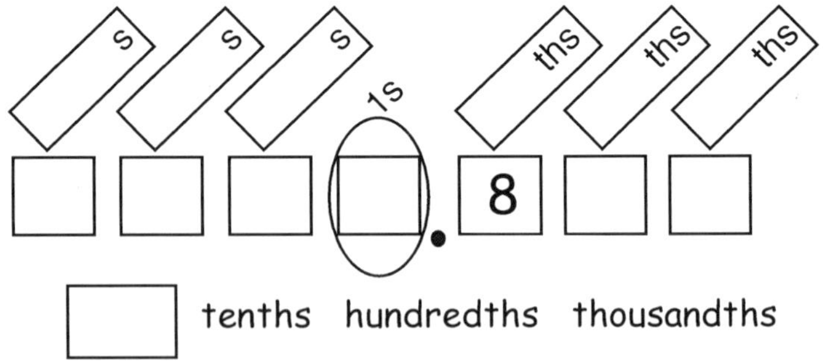

5. Write the number and circle the place of the last digit

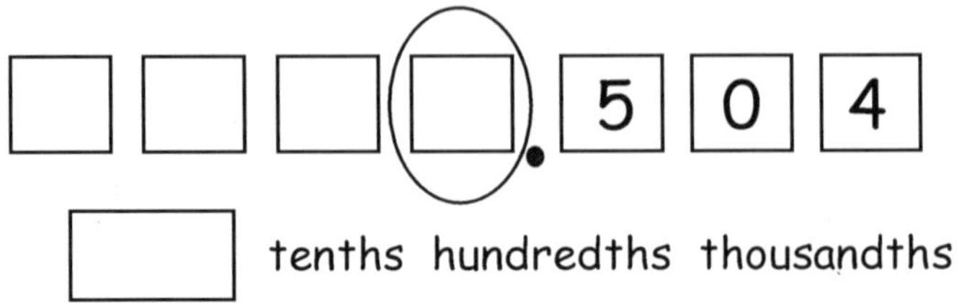

Concept Quiz

1. Write the decimal amount shaded

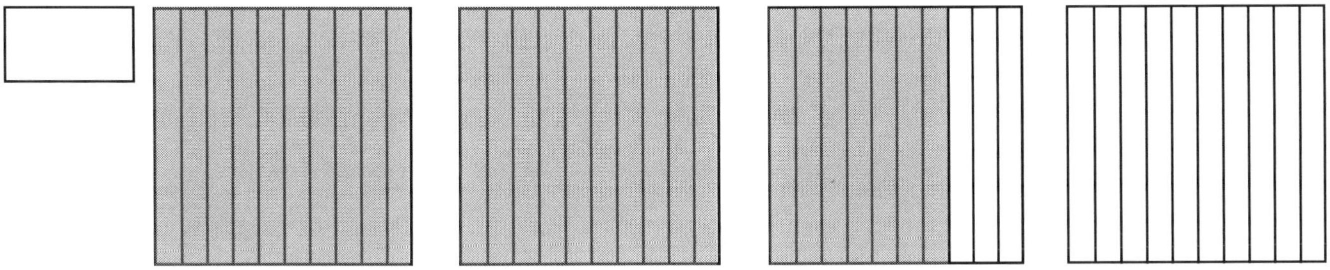

2. 3.6 Shade in the correct amount

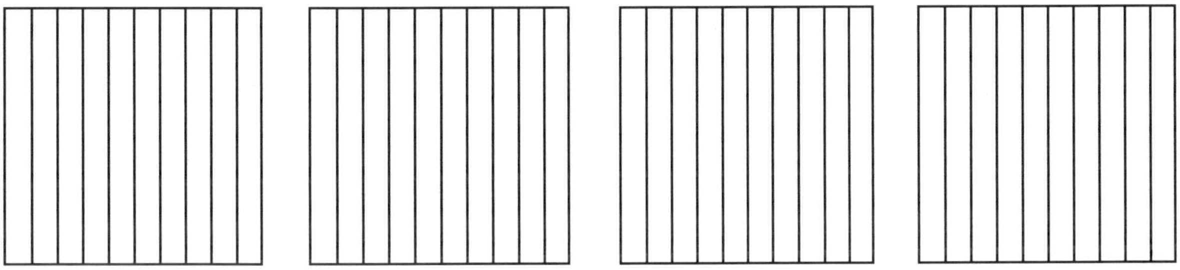

3.

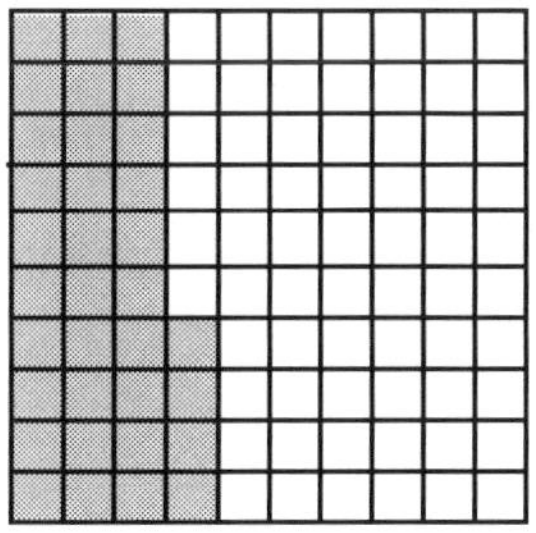

Decimal amount shaded: ______

Decimal amount UNshaded: ______

4. Label the boxes with the correct place value and draw rainbow lines. Show the multiplication associated with each place value box.

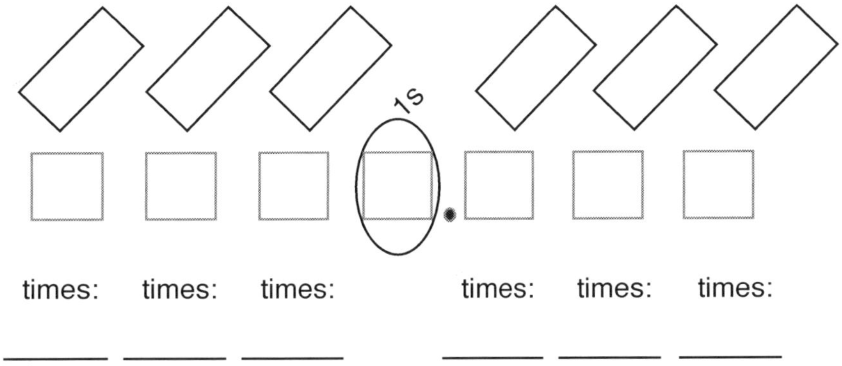

Naming Decimals

Put the number in the box and then circle the correct name

REMEMBER: "LAST DIGIT UP"!

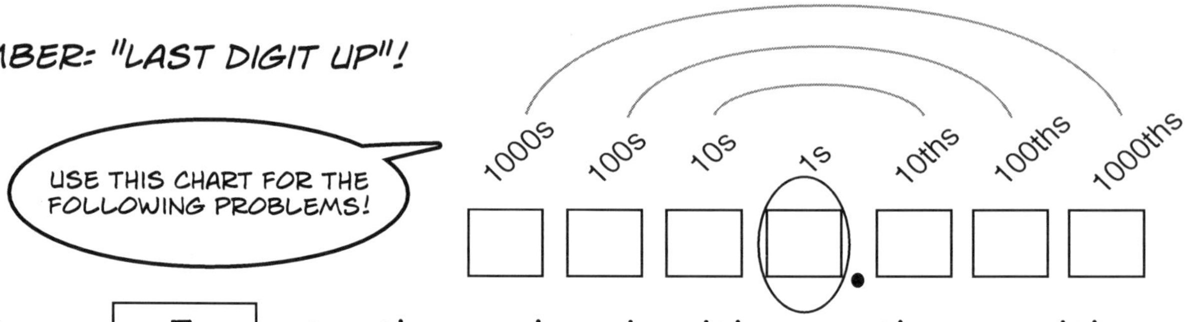

1.	.5	5	tenths	hundredths	thousandths
2.	.04		tenths	hundredths	thousandths
3.	.27		tenths	hundredths	thousandths
4.	.3		tenths	hundredths	thousandths
5.	.685		tenths	hundredths	thousandths
6.	.81		tenths	hundredths	thousandths
7.	.008		tenths	hundredths	thousandths
8.	.046		tenths	hundredths	thousandths
9.	.9		tenths	hundredths	thousandths
10.	.937		tenths	hundredths	thousandths

Writing Decimals

Put the decimal digits in the correct boxes

1. 25 hundredths

1000s	100s	10s	1s	10ths	100ths	1000ths
			.			

2. 25 thousandths

1000s	100s	10s	1s	10ths	100ths	1000ths
			.			

DON'T FORGET ZERO AS A PLACEHOLDER! —YOU CAN'T HAVE BLANKS!

3. 379 thousandths

1000s	100s	10s	1s	10ths	100ths	1000ths
			.			

4. 17 hundredths

1000s	100s	10s	1s	10ths	100ths	1000ths
			.			

5. 58 thousandths

1000s	100s	10s	1s	10ths	100ths	1000ths
			.			

6. 906 thousandths

1000s	100s	10s	1s	10ths	100ths	1000ths
			.			

"And" Means Decimal Point

Examples

LEFT OF THE DECIMAL ← → RIGHT OF THE DECIMAL

A. 4 and 7 tenths

WHOLE NUMBER PART ← ↑ → DECIMAL (FRACTION) PART

"AND" MEANS DECIMAL POINT!

4 . 7

B. 23 and 84 hundredths

whole number — decimal point — decimal part

2 3 . 8 4

Structure of Whole Number & Decimal Numbers

whole number . decimal part

↑ "and"

Put the decimal digits in the correct boxes

1. 6 and 4 tenths

1000s	100s	10s	1s	.	10ths	100ths	1000ths
				.			

2. 9 and 37 thousandths

1000s	100s	10s	1s	.	10ths	100ths	1000ths
				.			

3. 89 and 71 hundredths

1000s	100s	10s	1s	.	10ths	100ths	1000ths
				.			

Writing Decimals

Now put digits on both sides of the decimal point

Put the decimal digits in the correct boxes

1. 17 and 2 tenths

1000s	100s	10s	1s	.	10ths	100ths	1000ths
				.			

2. 28 and 13 thousandths

1000s	100s	10s	1s	.	10ths	100ths	1000ths
				.			

3. 89 and 71 hundredths

1000s	100s	10s	1s	.	10ths	100ths	1000ths
				.			

4. 4 and 506 thousandths

1000s	100s	10s	1s	.	10ths	100ths	1000ths
				.			

5. 605 and 27 thousandths

1000s	100s	10s	1s	.	10ths	100ths	1000ths
				.			

6. 76 and 49 hundredths

1000s	100s	10s	1s	.	10ths	100ths	1000ths
				.			

Concept Quiz

1. Put the number in the box and then circle the correct name

.237 ☐ tenths hundredths thousandths

2. Put the digits in the correct boxes

1000s 100s 10s 1s 10ths 100ths 1000ths

48 hundredths → ☐ ☐ ☐ ☐ . ☐ ☐ ☐

3. ☐ ☐ ☐ ☐ . ☐ ☐ ☐

↑

Every number in this box is really multiplied by: ☐

4. Write the number "9 and 2 tenths" ☐ ☐ ☐ ☐ . ☐ ☐ ☐

5. Write the number "7 and 5 hundredths" ☐ ☐ ☐ ☐ . ☐ ☐ ☐

6. Write the number "34 and 207 thousandths" ☐ ☐ ☐ ☐ . ☐ ☐ ☐

The DM-10 Trick

D – DIVIDE
M – MULTIPLY
10 – WITH TEN OR POWERS OF TEN (100'S, 1000S, 10THS, 100THS, ETC.)

IF YOU MULTIPLY OR DIVIDE A NUMBER BY A POWER OF 10, YOU JUST SLIDE THE DECIMAL ON THE ORIGINAL NUMBER!

D (DIVIDE)	M (MULTIPLY)
←	→
When you divide by 10s you slide to the LEFT	**When you multiply by 10s you slide to the RIGHT**

THIS IS WHY THE "D" COMES FIRST IN "DM-10"!

LET'S START BY DIVIDING BY 10S

WHEN YOU DIVIDE, YOU SLIDE TO THE LEFT!

Examples

A. $5.92 \div 10 =$ .592

ONE ZERO, ONE SLIDE!

B. $.98 \div 10 =$ .098

IF YOU NEED TO, ADD ONE OR MORE ZEROS TO GIVE ROOM TO SLIDE

(YOU CAN'T SLIDE IN BLANK SPACE!)

1. $62 \div 10 =$ ____

REMEMBER! THERE IS AN INVISIBLE DECIMAL POINT HERE!

2. $4 \div 10 =$ ____

3. $.7 \div 10 =$ ____

4. $700 \div 10 =$ ____

5. $700 \div 100 =$ ____

TWO ZEROS, TWO SLIDES!

6. $84.5 \div 100 =$ ____

7. $4 \div 100 =$ ____

8. $937 \div 1{,}000 =$ ____

STILL TO THE LEFT BECAUSE WE ARE DIVIDING!

HOW MANY SLIDES DO YOU THINK WITH THREE ZEROS?

9. $214.8 \div 1{,}000 =$ ____

TIP: IT HELPS TO DRAW ARROWS FOR EACH SLIDE!

The DM-10 Trick with Multiplication

D (DIVIDE)	M (MULTIPLY)
←	→
When you divide by 10s you slide to the LEFT	When you multiply by 10s you slide to the RIGHT

LET'S START BY MULTIPLYING BY 10S

WHEN YOU MULTIPLY, YOU SLIDE TO THE RIGHT!

Examples

A. 2.46 × 10 = 24.6

ONE ZERO, ONE SLIDE!

B. 15 × 10 = 150

other ways of writing 15 → 15. → 15.0

1. .65 · 10 = ____

2. .872 · 100 = ____

3. .91 · 100 = ____

4. .2 · 100 = ____

5. 30 · 10 = ____

6. .07 · 10 = ____

7. .01 · 100 = ____

8. .0053 · 100 = ____

9. .04 · 1,000 = ____

10. 7.52 · 1,000 = ____

11. 29.2 · 100 = ____

12. 813.4 · 100 = ____

The DM-10 Trick with Multiplication & Division

D (DIVIDE) ←
When you divide by 10s you slide to the LEFT

M (MULTIPLY) →
When you multiply by 10s you slide to the RIGHT

1.	$2.87 \cdot 10 =$	**10.**	$.003 \cdot 10 =$
2.	$36.2 \div 10 =$	**11.**	$.003 \cdot 100 =$
3.	$365 \div 100 =$	**12.**	$.01 \cdot 100 =$
4.	$.86 \div 10 =$	**13.**	$7 \div 100 =$
5.	$4.72 \cdot 10 =$	**14.**	$.17 \cdot 1000 =$
6.	$3.5 \cdot 100 =$	**15.**	$781 \div 10 =$
7.	$3.5 \div 100 =$	**16.**	$.047 \cdot 100 =$
8.	$.009 \cdot 100 =$	**17.**	$8.94 \cdot 10 =$
9.	$.1 \cdot 10 =$	**18.**	$.2 \div 100 =$

Seeing Decimals as Division

SHOW THE DIVISION PROBLEM INDICATED BY EACH DECIMAL

CIRCLE THE CORRECT ANSWER!

1.	.3 = 3 ÷	10	100	1000
2.	.1 = 1 ÷	10	100	1000
3.	.01 = 1 ÷	10	100	1000
4.	.52 = 52 ÷	10	100	1000
5.	.347 = 347 ÷	10	100	1000
6.	.092 = 92 ÷	10	100	1000
7.	.25 = 25 ÷	10	100	1000
8.	.602 = 602 ÷	10	100	1000
9.	.06 = 6 ÷	10	100	1000
10.	.007 = 7 ÷	10	100	1000
11.	.4 = 4 ÷	10	100	1000

WRITE THE DECIMALS:

12.	4 ÷ 10	
13.	4 ÷ 100	
14.	27 ÷ 10	
15.	27 ÷ 100	
16.	419 ÷ 1000	
17.	6 ÷ 100	
18.	12 ÷ 100	
19.	5 ÷ 1000	
20.	19 ÷ 1000	
21.	2 ÷ 10	
22.	425 ÷ 1000	

Decimals as Dividing by Multiples of Ten

WHEN YOU DIVIDE, YOU SLIDE TO THE LEFT! (ONE SLIDE FOR EACH ZERO!)

THIS IS SIMILAR TO THE "DM-10 TRICK"!

A. $.6 = 6 \div \boxed{10}$ or $6 \cdot \frac{\boxed{1}}{\boxed{10}}$

THE BOTTOM NUMBERS OF FRACTIONS HAVE THE IDEA OF DIVISION!

Write each decimal as a division and fraction problem

1. $.03 = 3 \div \boxed{}$ or $3 \cdot \frac{\boxed{1}}{\boxed{}}$

2. $.42 = 42 \div \boxed{}$ or $42 \cdot \frac{\boxed{}}{\boxed{}}$

3. $.002 = 2 \div \boxed{}$ or $2 \cdot \frac{\boxed{}}{\boxed{}}$

4. $.039 = 39 \div \boxed{}$ or $39 \cdot \frac{\boxed{}}{\boxed{}}$

5. $.364 = 364 \div \boxed{}$ or $364 \cdot \frac{\boxed{}}{\boxed{}}$

6. $.704 = \boxed{} \div \boxed{}$ or $\boxed{} \cdot \frac{\boxed{}}{\boxed{}}$

Insights into Decimals

The number of decimal slides match the zeros on the bottom of a fraction

Examples

A.

$.1 = \frac{1}{10}$

one slide... one zero

$.3 = \frac{3}{10}$

one slide... one zero

B.

$.01 = \frac{1}{100}$

two slides... two zeros

$.07 = \frac{7}{100}$

two slides... two zeros

C.

$.001 = \frac{1}{1000}$

three slides... three zeros

$.004 = \frac{4}{1000}$

three slides... three zeros

NOW YOUR TURN TO PRACTICE:

1. Write .06 as a fraction $\frac{\square}{1\square}$

2. Write $\frac{9}{1000}$ as a decimal $\square$

D. **Example**

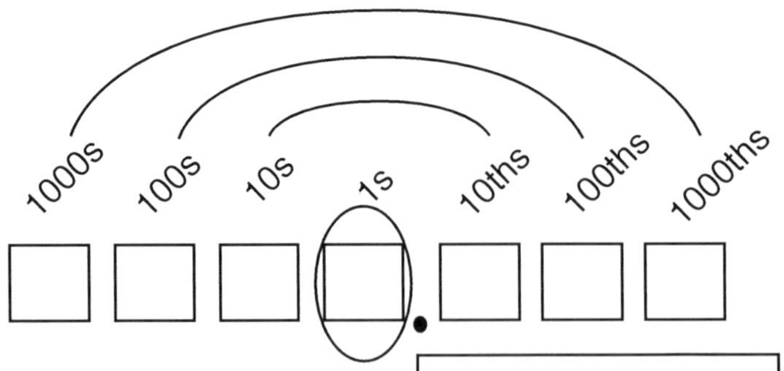

× 1000 × 100 × 10 × 1 $\times \frac{1}{10}$ $\times \frac{1}{100}$ $\times \frac{1}{1000}$

Same as what we had before: × .1 × .01 × .001

THIS IS ANOTHER WAY TO SHOW THE PLACE VALUE OF THE DECIMAL NUMBERS!

Decimal is **1 slide** to the left of the digit in this place	Decimal is **2 slides** to the left of the digit in this place	Decimal is **3 slides** to the left of the digit in this place

The DZ ("Dizzy") Trick

How to turn decimals into fractions

THE "D-Z TRICK" PUT A ***ZERO*** UNDER EVERY ***DECIMAL DIGIT!***

THEN CHANGE THE DECIMAL POINT TO A NUMBER ONE

AND ADD A FRACTION LINE! YOU'RE DONE!

IT WORKS SO FAST, IT MAY MAKE YOU FEEL D-Z ("DIZZY")!

THREE NUMBERS TO THE RIGHT OF THE DECIMAL!

A. 583 → 1000

TURN THE DECIMAL INTO A 1!

WRITE A ZERO UNDER EACH DECIMAL NUMBER!

$= \frac{583}{1000}$

NUMBERS TO THE ***LEFT*** OF THE DECIMAL POINT ARE ***WHOLE NUMBERS***--BIG NUMBERS BEFORE THE FRACTION!

B. $2.918 = 2\frac{918}{1000}$

000

- **#1** Put a zero under every decimal digit
- **#2** Trade the decimal point for a 1 to the LEFT of the zeros
- **#3** Draw a fraction bar

Write the following decimals as fractions

1. $.467 = \frac{\square}{1\square}$

000

2. $.742 = \frac{\square}{1\square}$

000

3. $.61 = \frac{\square}{1\square}$

4. $3.792 = \square\frac{\square}{1\square}$

5. $65.83 = \square\frac{\square}{1\square}$

6. $87.569 = \square\frac{\square}{1\square}$

7. $7.2836 = \square\frac{\square}{1\square}$

The DP Trick

DR. DECIMAL

2 slides to the **LEFT**, if you have a PERCENT and want to make a DECIMAL

It takes 2 slides

⟷ ⟷

to get from one side to another!

PEPPER PERCENT

2 slides to the **RIGHT**, if you have a DECIMAL and want to make a PERCENT

Use the DP Trick to change percents to decimals and the other way around

	DR. DECIMAL		PEPPER PERCENT
A.	.60 (or .6)	= slide the decimal point 2 times to the LEFT	60% (60.%)
1.	☐	slide the decimal point 2 times to the LEFT	40%
2.	☐	slide the decimal point 2 times to the LEFT	35%
3.	☐	slide the decimal point 2 times to the LEFT	7% (07.%)
4.	☐	slide the decimal point 2 times to the LEFT	467%

THE INVISIBLE DECIMAL POINT STARTS HERE!

YOU NEED TO ADD A ZERO BEFORE THE 6 TO HAVE ROOM TO SLIDE TWO TIMES!

5.	.20	slide the decimal point 2 times to the RIGHT	☐ %
6.	.84	slide the decimal point 2 times to the RIGHT	☐ %
7.	.3 (= .30)	slide the decimal point 2 times to the RIGHT	☐ %
8.	.05	slide the decimal point 2 times to the RIGHT	☐ %
9.	6.31	slide the decimal point 2 times to the RIGHT	☐ %

The DP Trick

Just slide two times to go from decimals to percents (or percents to decimals)

Use the DP Trick to change percents to decimals or the other way around

	DR. DECIMAL		PEPPER PERCENT				
1.	.48	=	☐ %	10.	3.2	=	☐ %
2.	☐	=	10%	11.	☐	=	.7%
3.	☐	=	38%	12.	☐	=	125%
4.	.05	=	☐ %	13.	.003	=	☐ %
5.	.5	=	☐ %	14.	15	=	☐ %
6.	5	=	☐ %	15.	☐	=	18.6%
7.	☐	=	971%	16.	☐	=	63.7%
8.	☐	=	8%	17.	4.19	=	☐ %
9.	.049	=	☐ %	18.	☐	=	400%

Decimals on a Number Line

A.

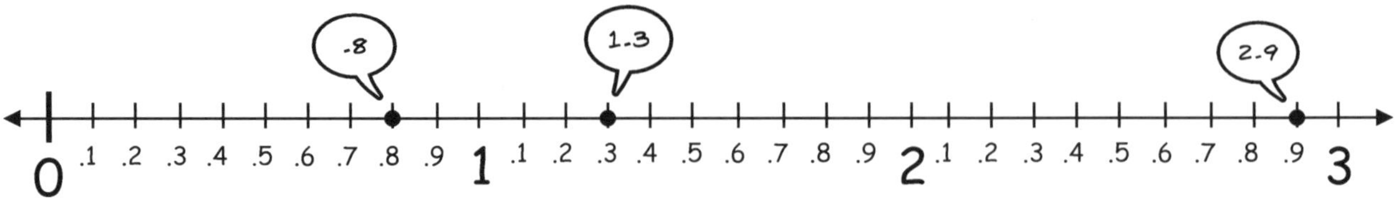

Write the decimal numbers for each of the points on the line

1.

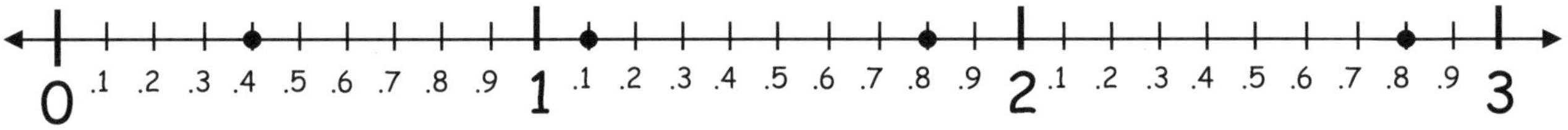

_____ _____ _____ _____

2.

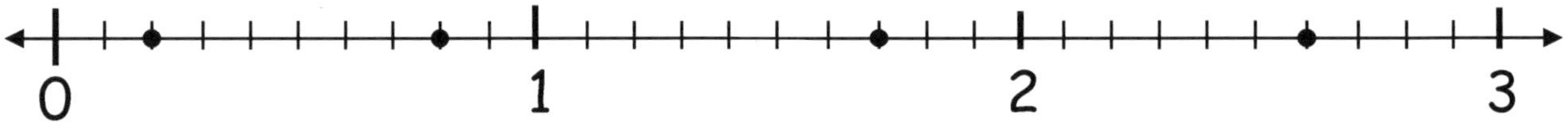

_____ _____ _____ _____

DRAW and LABEL points on the number line for the following decimals

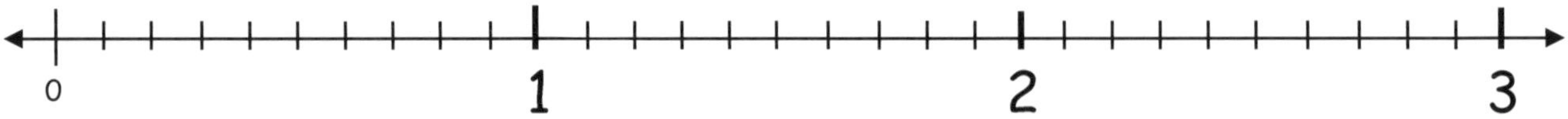

3. .5 **4.** .8 **5.** 1.2 **6.** 1.9 **7.** 2.4

Continue counting by TENTHS

8. 1.1 1.2 _____ 1.4 _____ _____ _____ _____ _____ _____ _____

Number Triangles with Decimals

Example

A. .4 3 9 8

.4
+ .0 3
+ .0 0 9
+ .0 0 0 8

HERE IS THE ORIGINAL NUMBER!

Number triangles help to show expanded notation. They enable you to see how decimal numbers are constructed!

Make number triangles like the example above

DO THESE THE EXACT SAME WAY. IGNORE THE BOXES THAT YOU DON'T NEED!

1. .8 3 4 2

.
+ .
+ .
+ .

2. .4 5 7 9

.
+ .
+ .
+ .

3. .6 3 1 5

.
+ .
+ .
+ .

4. .3 3 0 8

.
+ .
+ .
+ .

Expanded Notation to Standard Notation

Examples

In order (greatest to least)

A. .7 + .0 2 + .0 0 5 + .0 0 0 6 = .7 2 5 6

.7
+ .0 2
+ .0 0 5
+ .0 0 0 6
= .7 2 5 6

"Expanded Notation" "Standard Notation"

Not in order

B. .0 0 4 + .9 + .0 0 0 1 + .0 3 = .9 3 4 1

Convert the numbers from EXPANDED NOTATION to STANDARD NOTATION

1. .0 0 8 + .5 + .0 0 0 4 + .0 3 =

PUT THE DECIMALS IN THE RIGHT PLACES!

2. .0 0 0 7 + .0 1 + .6 + .0 0 9 =

3. .0 0 6 + .7 + .0 0 0 2 =

4. .0 3 + .0 0 8 + .0 0 0 9 + .2 =

Decimals Practice

1. .65 · 10 = ☐
(DM-10 Trick)

2. .31 · 1000 = ☐
(DM-10 Trick)

3. 214 ÷ 100 = ☐
(DM-10 Trick)

4. 8.7 ÷ 100 = ☐
(DM-10 Trick)

5. .29 = ☐/☐
(DZ Trick)

6. .071 = ☐/☐
(DZ Trick)

7. .3 = ☐/☐
(DZ Trick)

8. 4.897 = ☐ ☐/☐
(DZ Trick)

9. .3 = 3 ÷ ☐

10. .75 = 75 ÷ ☐

11. .45 = ☐ %

12. ☐ = 7 %

13. .063 = ☐ %

14. ☐ = 2.8 %

15. .94 = 94 · ☐/☐

16. .127 = 127 · ☐/☐

17. Write this decimal in expanded notation using the Number Triangle Trick

.5193

.☐
\+ .☐☐
\+ .☐☐☐
\+ .☐☐☐☐

Decimals Practice

1. $.5 = \frac{5}{\square}$

2. $\frac{7}{10} = \square$

3. $.43 = \frac{43}{\square}$

NOTICE THAT THE NUMBER OF ZEROS MATCHES THE NUMBER OF DECIMAL SLIDES!

4. $.675 = \frac{\square}{\square}$

5. $\frac{82}{100} = \square$

6. $.917 = \frac{\square}{\square}$

7. $.028 = \frac{\square}{\square}$

8. $\frac{48}{100} = \square$

9. $\frac{37}{1000} = \square$

10. $.917 = \frac{\square}{\square}$

11. Put into standard notation (regular decimal number)

$.3 + .01 + .002 + .0005$

$\square$

12. Put into standard notation (regular decimal number)

$.009 + .5 + .0006 + .08$

$\square$

13. Write the following decimal in expanded notation

$.6472$

$\square + \square + \square + \square$

14. Write 1.618 as a percent:

$\square$ %

Adding and Subtracting Elevator

Examples

A. 2.4 + .78

```
  2.4
+  .78
 ----
  3.18
```

WHENEVER YOU ADD OR SUBTRACT DECIMALS YOU NEED A DECIMAL ELEVATOR!

B. 7.8 - .46

BLANK SPACE MINUS 6??

```
  7.8
-  .46
 ----
```

YOU CAN'T HAVE BLANKS WHEN YOU SUBTRACT!

ADD A ZERO SO YOU CAN REGROUP AND THEN SUBTRACT!

```
     7 1
  7.8 0
-  .4 6
 ------
  7.3 4
```

PUT THE DIGITS IN THE CORRECT BOXES!

NOT ALL BOXES WILL BE USED!

ADD ZEROS WHERE THEY BELONG!

1. 5.6 + .37

```
   [ ] . [ ] [0]
+  [0] . [ ] [ ]
   -------------
   [ ] . [ ] [ ]
```

WRITE ZEROS WHENEVER YOU HAVE BLANKS, ESPECIALLY WHENEVER YOU SUBTRACT DECIMALS!

2. 2.7 - .13

```
   [ ] . [ ] [0]
-  [0] . [ ] [ ]
   -------------
   [ ] . [ ] [ ]
```

3. 3.66 + 5.4

```
   [ ] . [ ] [ ]
+  [ ] . [ ] [0]
   -------------
   [ ] . [ ] [ ]
```

4. 8.45 - 7.6

```
   [ ] . [ ] [ ]
-  [ ] . [ ] [ ]
   -------------
   [ ] . [ ] [ ]
```

5. 5.4 - 3.44

```
   [ ] . [ ] [ ]
-  [ ] . [ ] [ ]
   -------------
   [ ] . [ ] [ ]
```

6. 5 - 2.71

```
   [ ] . [ ] [ ]
-  [ ] . [ ] [ ]
   -------------
   [ ] . [ ] [ ]
```

7. 7.03 - 4.9

```
   [ ] . [ ] [ ]
-  [ ] . [ ] [ ]
   -------------
   [ ] . [ ] [ ]
```

Adding and Subtracting Elevator

Line up the decimals and add or subtract; put zeros in blank boxes

You line up the decimals when you ADD or SUBTRACT; but NOT when you divide or multiply

1. .3 2 + 5.7

PUT ADDITION OR SUBTRACTION SIGNS HERE!

4. 3.4 1 + 2.0 7 1

2. 7.5 3 - .4 4

5. 8.3 4 - 4.0 5 2

DON'T FORGET TO ADD A ZERO! YOU CAN'T SUBTRACT WITH BLANKS!

3. 5.0 7 - .1 8

6. 5.6 - 2.0 3 7

Decimal Addition & Subtraction Practice

Line up the decimals, and add or subtract

Line up the numbers correctly before adding or subtracting

1. .2 4 + 1 .9

Answer: ☐ . ☐ ☐

4. 8.6 7 + .0 8 9 4

Answer: ☐

2. 8.0 7 - .6 5

Answer: ☐ . ☐ ☐

5. 9.0 2 - 3.0 4 7

Answer: ☐

3. 3.2 4 - .37

Answer: ☐ . ☐ ☐

6. 8.6 - 5.4 3 1

Answer: ☐

Adding Decimals Mentally

Line up the decimals MENTALLY; then add
Afterwards write the numbers in the boxes to check your answer

1. .3 +.5 = ☐

CHECK YOUR ANSWER IN THE BOXES AFTER DOING THE PROBLEM MENTALLY!

2. .7 +.7 = ☐

3. .0 6 + .1 7 = ☐

4. 0.7 1 + 0.6 = ☐

5. 1.6 + 0.3 5 = ☐

6. 2.6 4 + 2.6 = ☐

Adding Decimals Mentally

Line up the decimals MENTALLY; then add

THIS MAY HELP YOU TO VISUALIZE HOW TO LINE UP THE DECIMALS THAT YOU SOLVE MENTALLY ON THIS PAGE!

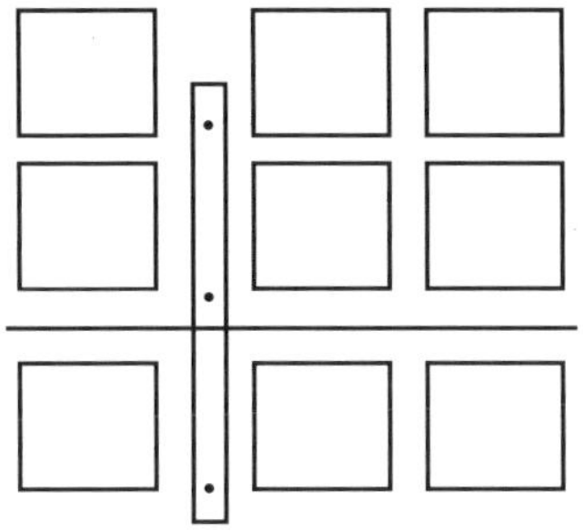

1. .8 + .08 =
2. 3.8 +.8 =
3. .7 + .03 =
4. .7 + .3 =
5. .6 + .25 =
6. .08 + .06 =
7. .18 + .06 =
8. .62 + .08 =
8. .02 + .08 =
9. 6.85 + 0.5 =
10. .03 + .17 =
11. .45 + .45 =
12. 3.6 + 3.6 =
13. 2.7 + 2.76 =
14. 49.8 + 0.5 =
15. 8.96 +.04 =
16. 7.35 +.36 =
17. 5.73 + 1.09 =
18. 4.86 +.17 =

Subtracting Decimals Mentally

Line up the decimals MENTALLY; then subtract.
Afterwards, write the numbers in the boxes to check your answer.

1. 1.2 - .6 = ☐

CHECK YOUR ANSWER IN THE BOXES AFTER DOING THE PROBLEM MENTALLY!

4. 2.57 - 0.3 = ☐

2. 1.0 - .3 = ☐

5. 4.25 - 0.3 = ☐

3. 20.0 - .7 = ☐

6. 5.4 - 0.25 = ☐

Subtracting Decimals Mentally

Line up the decimals MENTALLY; then subtract.
Afterwards write the numbers in the boxes to check your answer.

1. 4.8 - .7 = ☐

2. 1.0 - .6 = ☐

3. 1.4 - .7 = ☐

4. 40.0 - .7 = ☐

5. 8.57 - 0.3 = ☐

6. 3.21 - 0.3 = ☐

7. 5.4 - 0.05 = ☐

8. 5.4 - 0.25 = ☐

9. 3.38 - 0.2 = ☐

10. 7.6 - 0.38 = ☐

REVIEW: Line up the decimals MENTALLY, then add

11. 2.46 + .3 = ☐

12. 2.46 +.02 = ☐

13. 2.46 +.32 = ☐

14. 5.57 +.06 = ☐

15. 3.83 +.80 = ☐

16. 3.76 +.75 = ☐

Decimal Addition-Subtraction Quiz

Add or subtract the following decimals, following all the steps you have learned so far

1. 7 - 3.62 = ______

4. 95.302 - 23.148 = ______

2. 7.06 + 35.8 = ______

5. 16.035 - 2.993 = ______

3. 8.5 - 2.07 = ______

6. 14.008 - 6.037 = ______

Whole Numbers Minus Decimals

Example

A. 4 - .06

Wrong!

```
  4
- .06
```

THERE IS NO SUCH THING AS "SPACE MINUS 6!"

YOU CAN'T HAVE BLANKS WHEN YOU SUBTRACT!

You have to subtract from zeros instead of blanks

Right!

```
  4.00
-  .06
```

ADD A DECIMAL POINT AND ZEROS ABOVE EVERY DIGIT YOU SUBTRACT FROM!

How to subtract when you have a whole number minus a decimal

#1 Add a decimal point to the right of any whole number

YOU CAN ADD AS MANY ZEROS TO THE ***RIGHT*** OF A DECIMAL AS YOU WANT!

#2 Line up the decimal points and add any placeholder zeros where needed

#3 Subtract normally—and bring down the decimal point

Subtract the following decimals from whole numbers

1. 8 - .4

```
  □.□□□
- □.□□□
```

2. 9 - .76

```
  □.□□□
- □.□□□
```

3. 3 - .185

```
  □.□□□
- □.□□□
```

4. 17 - .26

```
  □□.□□□
- □□.□□□
```

Concept Quiz 1

1. Tenths are represented by

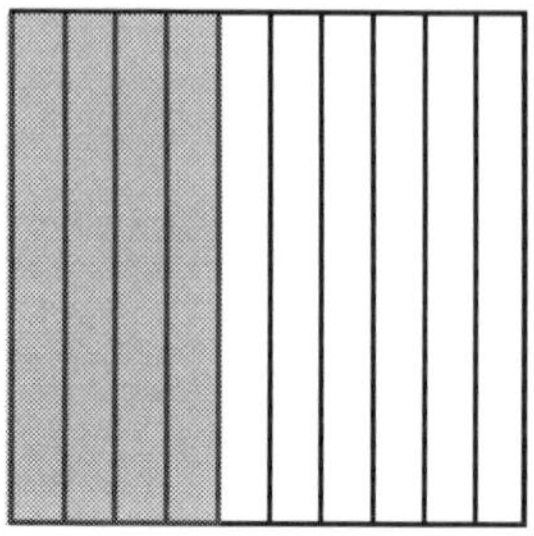

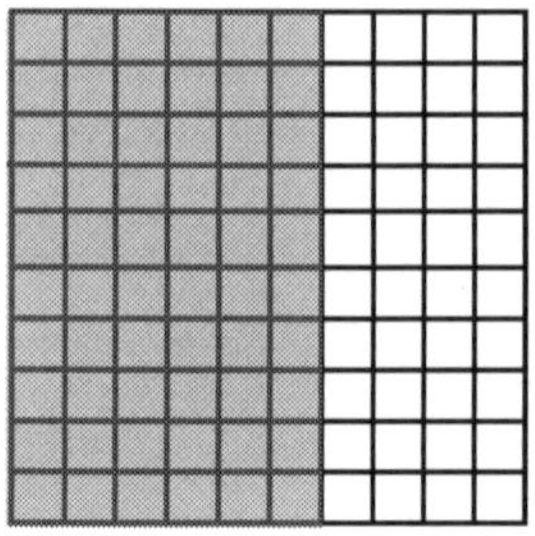

LONG BARS (Circle one) SMALL BOXES

2. Label the boxes with the correct place value and draw rainbow lines.

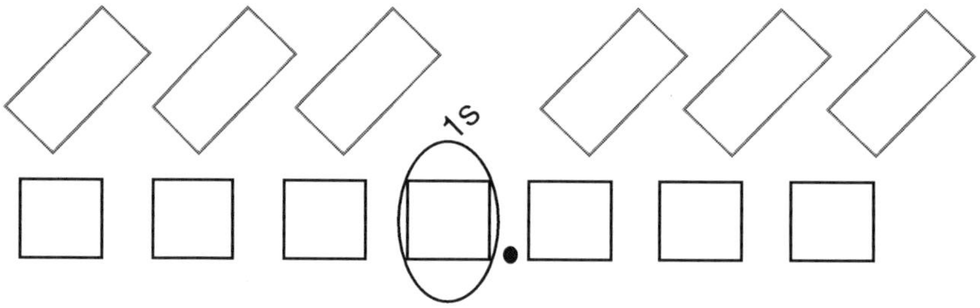

times: ______ times: ______ times: ______ times: ______ times: ______ times: ______

3.

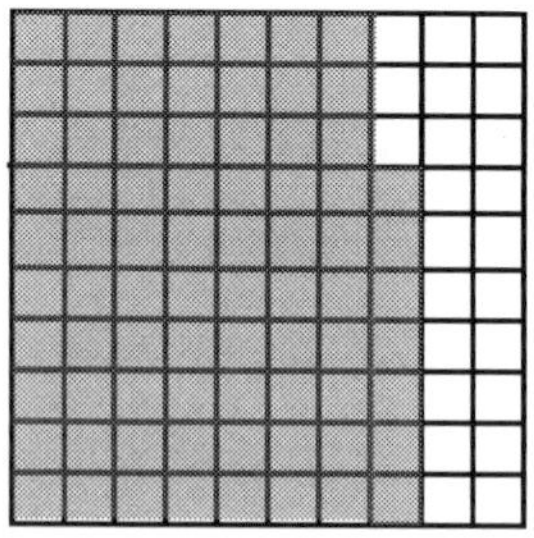

WRITE YOUR ANSWERS AS ***DECIMALS!***

Decimal amount shaded: [ ]

Decimal amount UNshaded: [ ]

4. When reading decimal numbers, the decimal point represents this word: ____________

5. Write 24 and 79 thousandths:

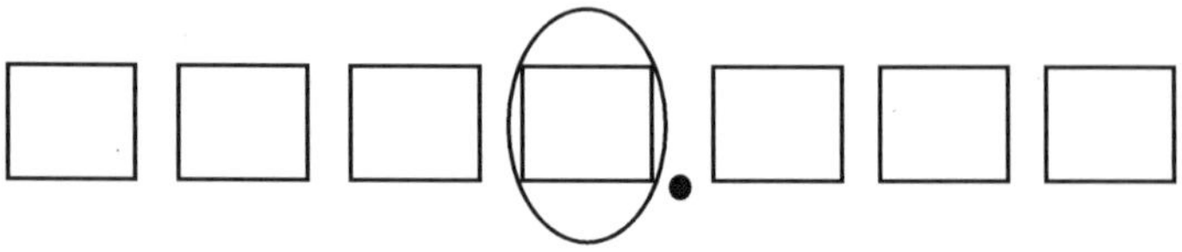

Concept Quiz 2

1. What are the three steps you need to take when adding or subtracting decimals (like 7 - .23)?

Step 1: ______________________________

Step 2: ______________________________

Step 3: ______________________________

2. What is wrong with the following problem?

$$\begin{array}{r} .75 \\ -\quad .3 \\ \hline \end{array}$$

____

____

3. What is missing in the following problem?

$$\begin{array}{r} 4.2 \\ -\ 2.38 \\ \hline \end{array}$$

____

____

4. Write the following decimals as fractions or mixed numbers:

(a) .859 ☐ (b) 2.37 ☐

Show all the steps as you solve these decimal problems:

5. 6.3 - 2.07 =

6. 7.05 + 5.89 =

Draw and label points on the number line for the following decimals:

6. .7 **7.** 1.3 **8.** 2.1 **9.** 2.9

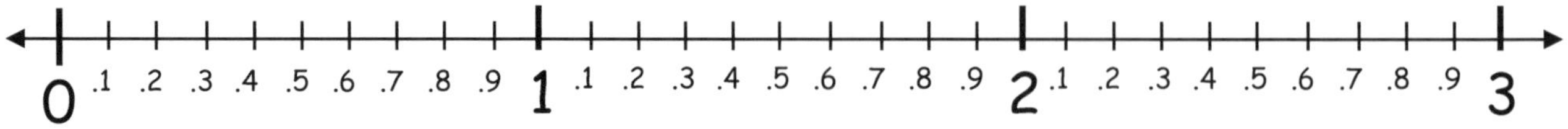

10. Make a number triangle to show expanded notation

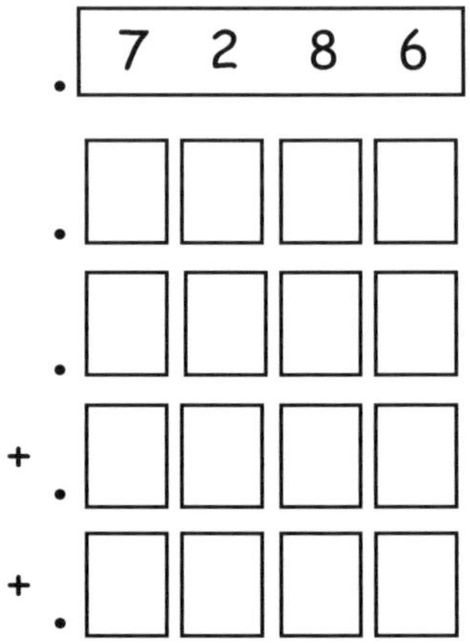

11. Subtract the following decimals 7.1 - 3.83

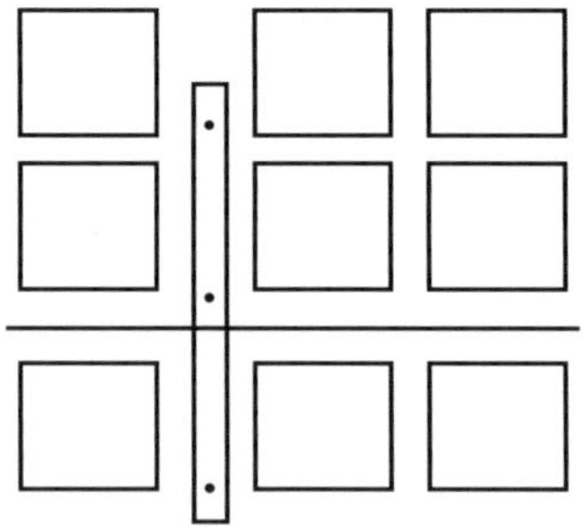

12. The Dizzy Trick - Turn .942 into a fraction

<u>Step one</u>: Write a _______ under each decimal digit

<u>Step two:</u> Change the decimal point to ________________

<u>Step three:</u> Add a ______________________ bar

Answer: [ ]

Decimals as Added Fractions

Example

$$.376 = \frac{3}{10} + \frac{7}{100} + \frac{6}{1000}$$

Write the following decimals as the sum of fractions (as above)

1. .52

$$\frac{\square}{10} + \frac{\square}{100}$$

2. .068

$$\frac{\square}{10} + \frac{\square}{100} + \frac{\square}{1000}$$

3. .709

$$\frac{\square}{10} + \frac{\square}{100} + \frac{\square}{1000}$$

4. .9837

$$\frac{\square}{10} + \frac{\square}{100} + \frac{\square}{1000} + \frac{\square}{10000}$$

5. .13

$$\frac{\square}{\square} + \frac{\square}{\square}$$

6. .003

$$\frac{\square}{\square} + \frac{\square}{\square} + \frac{\square}{\square}$$

7. .176

$$\frac{\square}{\square} + \frac{\square}{\square} + \frac{\square}{\square}$$

8. .042

$$\frac{\square}{\square} + \frac{\square}{\square} + \frac{\square}{\square}$$

Decimals as Added Fractions

Write the following decimals from the sum of fractions

1. .0 6

$\frac{\square}{10} + \frac{\square}{100}$

2. .2 7 4

$\frac{\square}{\square} + \frac{\square}{\square} + \frac{\square}{\square}$

3. $.\overline{3}$ = .333...

THE LINE ABOVE THE 3 MEANS THAT THE 3 REPEATS TO INFINITY!

$\frac{\square}{\square} + \frac{\square}{\square} + \frac{\square}{\square} + \cdots$

BACKWARDS!
NOW YOU WRITE THE DECIMAL!

4. $\frac{4}{10} + \frac{3}{100}$

$.\square\square$

WATCH WHERE YOU PUT THE DECIMAL NUMBERS!

THESE FRACTIONS ARE NOT PUT IN ORDER!

5. $\frac{7}{100} + \frac{2}{10}$

$.\square\square$

6. $\frac{6}{10} + \frac{5}{100} + \frac{2}{1000}$

$.\square\square\square$

7. $\frac{8}{1000} + \frac{1}{100} + \frac{3}{10}$

$.\square\square\square$

8. $\frac{4}{100} + \frac{9}{10} + \frac{7}{1000}$

$.\square\square\square$

9. $\frac{4}{10} + \frac{1}{1000} + \frac{7}{100}$

$.\square\square\square$

10. $\frac{0}{100} + \frac{6}{1000} + \frac{5}{10}$

$.\square\square\square$

Putting Decimals in Order

Example

A. Put the following decimals in order: .1, .009, .08

.	1		
.	0	0	9
.	0	8	

← 9 is normally larger than 1, but 1 in the tenths place is larger than 9 in the thousandths place!

↑ Even though 1 is smaller than 9 or 8, it is larger than zero! Placement matters!

Largest decimal: ______

Middle decimal: ______

Smallest decimal: ______

Line up the decimals and put them in order

1. .1 .01 .11

FIRST, COPY THE NUMBERS IN THE ORDER YOU SEE THEM IN THE PROBLEM!

Largest decimal: ______

Middle decimal: ______

Smallest decimal: ______

2. .65 .56 .66

Largest decimal: ______

Middle decimal: ______

Smallest decimal: ______

Putting Decimals in Order

Line up the decimals and put them in order from GREATEST to LEAST

THIS IS ACTUALLY VERY SIMILAR TO PUTTING WORDS IN ALPHABETICAL ORDER!

Check this column first | Check this column second | Check this column third

1. .09 .2 .089

. ☐☐☐ Largest decimal: ☐

. ☐☐☐ Middle decimal: ☐

. ☐☐☐ Smallest decimal: ☐

2. .07 .777 .7

. ☐☐☐ Largest decimal: ☐

. ☐☐☐ Middle decimal: ☐

. ☐☐☐ Smallest decimal: ☐

3. .181 .180 .118

. ☐☐☐ Largest decimal: ☐

. ☐☐☐ Middle decimal: ☐

. ☐☐☐ Smallest decimal: ☐

4. .207 .270 .072

. ☐☐☐ Largest decimal: ☐

. ☐☐☐ Middle decimal: ☐

. ☐☐☐ Smallest decimal: ☐

Ordering from LEAST to GREATEST

Line up the decimals and put them in order, from LEAST to GREATEST

1. .818 .881 .118

HERE, JUST PUT THE NUMBERS IN THE ORDER THEY ARE GIVEN!

.☐☐☐	Smallest decimal:	☐
.☐☐☐	Middle decimal:	☐
.☐☐☐	Largest decimal:	☐

2. .068 .086 .806

.☐☐☐	Smallest decimal:	☐
.☐☐☐	Middle decimal:	☐
.☐☐☐	Largest decimal:	☐

TIP: LABEL THE DECIMALS 1, 2, 3 (SMALLEST TO LARGEST)!

3. .321 .322 .312

.☐☐☐	Smallest decimal:	☐
.☐☐☐	Middle decimal:	☐
.☐☐☐	Largest decimal:	☐

4. .544 .455 .454

.☐☐☐	Smallest decimal:	☐
.☐☐☐	Middle decimal:	☐
.☐☐☐	Largest decimal:	☐

5. .518 .409 .589

.☐☐☐	Smallest decimal:	☐
.☐☐☐	Middle decimal:	☐
.☐☐☐	Largest decimal:	☐

Concept Quiz

1. Write the number "26 and 318 thousandths"

☐☐☐☐.☐☐☐

2. Put 9 in the hundredths place;
Put 5 in the tenths place;
Put 1 in the thousandths place;
Put 8 in the thousands place
Put zeros in any remaining places

☐☐☐☐.☐☐☐

3. $47 \div 10 =$ ☐

4. $2.51 \cdot 100 =$ ☐

5. Write the decimal as a sum of two fractions:

.67

$\frac{\square}{10} + \frac{\square}{100}$

6. Write the decimal from a sum of fractions:

$\frac{4}{1000} + \frac{5}{100} + \frac{2}{10}$.☐☐☐

7. Line up the decimals and put them and put them in order from least to greatest

.822 .882 .288

.☐☐☐ Smallest decimal: ☐

.☐☐☐ Middle decimal: ☐

.☐☐☐ Largest decimal: ☐

8. $3.5 - 1.02 =$ ☐

9. If you write this decimal as a fraction the bottom number will have a 1 with _______ zeros.

$.856 = \frac{\square}{1\square}$ ← How many zeros?

10. How do you know how many zeros go on the denominator in #9? (What is the rule?)

___

11. Whenever you DIVIDE by 10 or a power of 10 you slide the decimal to the

LEFT   RIGHT

12. $7.3 - 2.46 = \square$

13. $6.37 + .748 = \square$

14. Write the following decimal as a sum of fractions:

.976

$\frac{\square}{\square} + \frac{\square}{\square} + \frac{\square}{\square}$

15. Write the decimal form of this sum of factions

$\frac{2}{10} + \frac{5}{1000} + \frac{8}{100}$

$.\square\square\square$

16. $.001 = \frac{1}{1\square}$ ← $\square$ zeros

How many zeros will be on the bottom of the fraction?

17. $.004 + .2 + .0003$

$= \square$

Multiplying Decimals

Example

TWO SLIDES IN THE PROBLEM... TWO SLIDES IN THE ANSWER!

A. .6 × .3 = .18

COUNT THE TOTAL NUMBER OF SLIDES IN THE PROBLEM... → 2

GIVE YOUR ANSWER THE SAME NUMBER OF SLIDES!

#1 Add up the number of slides in ALL factors

#2 Multiply normally

#3 Give your answer the same number of slides as in #1

YOU CAN JOT DOWN THE NUMBER OF SLIDES IN THE PROBLEM AND THEN ERASE THE DECIMAL POINTS!

...BUT REMEMBER TO GIVE YOUR ANSWER THE SAME NUMBER OF DECIMAL SLIDES!

YOU ONLY LINE THEM UP WHEN YOU ADD OR SUBTRACT THEM!

You don't line up the decimals when you multiply or divide them

1. .3 × .8

#1 Number of slides in the problem: ☐

#2 Multiply normally 3 × 8

#3 Give your answer the same number of slides: . ☐☐

.3 × .8 = ☐

2. .04 × .7

AFTER YOU COUNT THE SLIDES YOU CAN FORGET THEM WHILE YOU CRUNCH THE NUMBERS!

#1 Number of slides in the problem: ☐

#2 Multiply normally 4 × 7 = ☐

#3 Give your answer the same number of slides: . ☐☐☐

.04 × .7 = ☐

DID YOU MAKE ROOM FOR THE EXTRA SLIDE (PUT IN A ZERO)?

Multiplying Decimals

Review

- **#1** Add up the number of slides in ALL factors
- **#2** Multiply normally
- **#3** Give your answer the same number of slides as in the whole problem (same as #1)

SERIOUSLY! YOU CAN EVEN PRETEND IT ISN'T A DECIMAL PROBLEM (AT LEAST UNTIL YOU GET TO THE END)!

Multiply the following decimals by following the steps given

1. .4 × .6

#1 Number of slides in the problem: ☐

#2 Multiply normally ☐ × ☐ = ☐

#3 Give your answer the same number of slides: .4 × .6 = ☐

SLIDE LEFT!

2. 7 × .05

MULTIPLY JUST THE WHOLE NUMBERS!

#1 Number of slides in the problem: ☐

#2 Multiply normally ☐ × ☐ = ☐

#3 Give your answer the same number of slides: 7 × .05 = ☐

3. .03 × .02

#1 Number of slides in the problem: ☐

#2 Answer when you just multiply the whole numbers: = ☐

#3 Give your answer the same number of slides .03 × .02 = ☐

Multiplying Decimals

Multiply the non-zero numbers and give your answer the right number of slides

THE NUMBER OF SLIDES HAS TO MATCH!

	Problem	Answer	Number of slides
1.	$.3 \times 8 =$	2.4	1
2.	$.3 \times .8 =$		
3.	$7 \times .04 =$		
4.	$.05 \times 7 =$		
5.	$.002 \times 8 =$		
6.	$.04 \times 9 =$		
7.	$.8 \times 4 =$		
8.	$.6 \times .9 =$		
9.	$.9 \times 5 =$		
10.	$.7 \times .06 =$		
11.	$9 \times .07 =$		
11.	$.8 \times .8 =$		
12.	$.7^2 =$		
13.	$.2 \times 10 =$		
14.	$.3 \times 20 =$		
15.	$.9^2 =$		
16.	$.4 \times .30 =$		
17.	$1000 \times .009 =$		

CONSIDER BOTH FACTORS!

Decimal Multiplication Practice

Multiply the non-zero numbers and give your answer the right number of slides

THE NUMBER OF SLIDES HAS TO MATCH!

1. $3 \times 5 =$ ☐ Number of slides: ☐
2. $.3 \times 5 =$ ☐ Number of slides: ☐
3. $.3 \times .5 =$ ☐ Number of slides: ☐
4. $.02 \times 6 =$ ☐ Number of slides: ☐
5. $.01 \times 365 =$ ☐ Number of slides: ☐
6. $4 \times .07 =$ ☐ Number of slides: ☐
7. $6 \times .3 =$ ☐ Number of slides: ☐
8. $.2 \times .2 \times .2 =$ ☐ Number of slides: ☐
9. $.1 \times 10 =$ ☐ Number of slides: ☐
10. $.01 \times 100 =$ ☐ Number of slides: ☐
11. $.3 \times .02 =$ ☐ Number of slides: ☐
12. $.3 \times .3 =$ ☐ Number of slides: ☐
13. $.5^2 =$ ☐ Number of slides: ☐

HINT! -- WRITE THIS OUT AS MULTIPLICATION PROBLEM!

14. $.06 \times 100 =$ ☐ Number of slides: ☐
15. $.8 \times 30 =$ ☐ Number of slides: ☐
16. $.2^2 =$ ☐ Number of slides: ☐
17. $.1 \times .1 \times .1 \times .1 =$ ☐ Number of slides: ☐
18. $.03 \times .04 =$ ☐ Number of slides: ☐

Decimal Multiplication Practice

Multiply the decimals and give your answer the right number of slides

THE NUMBER OF SLIDES HAS TO MATCH!

WHEN YOU MULTIPLY BY ***LESS THAN ONE,*** YOU ARE REALLY ***DIVIDING!***

1. .2 × 4 = ☐

2. .2 × .4 = ☐

3. .03 × .2 = ☐

4. .001 × 7 = ☐

5. 5 × .3 = ☐

6. .1 × .1 = ☐

7. 8 × .07 = ☐

8. .2 × .9 = ☐

9. .06 × .1 = ☐

10. .3 × .09 = ☐

11. .7 × 4 = ☐

12. .6 × .02 = ☐

13. .09 × .02 = ☐

14. .3 × .12 = ☐

15. .05 × .2 = ☐

16. .7 × .07 = ☐

17. 6 × .7 = ☐

18. .08 × .02 = ☐

19. 3 × .09 = ☐

20. .2 × .2 × .3 = ☐

21. .2 × .3 × .04 = ☐

22. .3 × .3 × 6 = ☐

Decimal Multiplication Practice

Multiply the decimals and give your answer the right number of slides

When you multiply by LESS than ONE, you are really ________________

1. .3 × .7 =

2. .3 × 7 =

3. .8 × .3 =

4. .12 × .3 =

5. .04 × .2 =

6. .9 × .09 =

7. 5 × .006 =

8. 6 × .8 =

9. .08 × .02 =

10. .01 × .039 =

11. .4 × .006 =

12. .2 × .3 × 10 =

MULTIPLYING BY 10 MOVES THE DECIMAL ONE TIME TO THE RIGHT!

13. .3 × 6 ÷ 10 =

DIVIDING BY 10 MOVES THE DECIMAL ONE TIME TO THE LEFT!

14. .7 × .3 ÷ 10 =

15. .8 × .6 × 10 =

16. .4 × .4 ÷ 10 =

17. .5 × 9 × 10 =

18. .3 × .9 ÷ 10 =

19. .5 × .07 × 10 =

20. .08 × .02 × 100 =

Multiplying Decimals

Examples

A.

```
    7 3
×    .3   ← 1 slide
-------
  2 9.2   ← 1 slide
```

ONE SLIDE IN THE PROBLEM...

ONE SLIDE IN THE ANSWER!

#1 Count the number of slides in the problem

#2 Multiply normally

#3 Give your answer the same number of slides

B.

```
    .3 6   ← 2 slides
  × .2 7   ← 2 slides
  ------
   2 5 2
   8 2 0
  ------
 .1 0 7 2  ← 4 slides
```

IGNORE THE DECIMALS AS YOU WORK THE PROBLEM!

JUST REMEMBER TO GIVE YOUR ANSWER THE SAME NUMBER OF SLIDES AS IN THE PROBLEM!

1.

```
   4.6   ← 1 slide
 × 3.2   ← 1 slide
 -----
```

Multiply normally (ignore the decimals)

Give your answer 2 slides

PUT YOUR ANSWER (WITH DECIMAL SLIDES) HERE!

2.

```
   .6 4   ← 2 slides
 × 2.5    ← 1 slide
 -----
```

Multiply normally (ignore the decimals)

Give your answer ___ slides

3.

```
    7 4
 × .0 2   ← ___ slide(s)
 ------
```

Multiply normally (ignore the decimals)

Give your answer ___ slides

4.

```
   .0 7   ← ___ slide(s)
 ×   .3   ← ___ slide(s)
 ------
```

Multiply normally (ignore the decimals)

Give your answer ___ slides

5.

```
   2 8 4   ← ___ slide(s)
 ×  .0 4   ← ___ slide(s)
 -------
```

Multiply normally (ignore the decimals)

Give your answer ___ slides

6.

```
   1.7 6   ← ___ slide(s)
 ×  .2 3   ← ___ slide(s)
 -------
```

Multiply normally (ignore the decimals)

Give your answer ___ slides

Multiplying Decimals

1.

$$\begin{array}{r} .57 \\ \times\ 3.5 \\ \hline \end{array}$$

← ___ slide(s)

← ___ slide(s)

Multiply normally
(ignore the decimals)

Give your answer ___ slides

PUT YOUR ANSWER (WITH DECIMAL SLIDES) HERE!

2.

$$\begin{array}{r} .36 \\ \times\ .8 \\ \hline \end{array}$$

← ___ slide(s)

← ___ slide(s)

Multiply normally
(ignore the decimals)

Give your answer ___ slides

3.

$$\begin{array}{r} 7.58 \\ \times\ .26 \\ \hline \end{array}$$

← ___ slide(s)

← ___ slide(s)

Give your answer ___ slides

4.

$$\begin{array}{r} 83 \\ \times .03 \\ \hline \end{array}$$

← ___ slide(s)

← ___ slide(s)

Give your answer ___ slides

5.

$$\begin{array}{r} 2.3 \\ \times\ 6.1 \\ \hline \end{array}$$

← ___ slide(s)

← ___ slide(s)

Give your answer ___ slides

6.

$$\begin{array}{r} .48 \\ \times\ 6.7 \\ \hline \end{array}$$

← ___ slide(s)

← ___ slide(s)

Give your answer ___ slides

7.

$$\begin{array}{r} 631 \\ \times\ .24 \\ \hline \end{array}$$

← ___ slide(s)

← ___ slide(s)

Give your answer ___ slides

8.

$$\begin{array}{r} .59 \\ \times .43 \\ \hline \end{array}$$

← ___ slide(s)

← ___ slide(s)

Give your answer ___ slides

Multiplying Decimals

1.

$$\begin{array}{r} 14.6 \\ \times\ 4.2 \\ \hline \end{array}$$

14.6 ← ___ slide(s)
× 4.2 ← ___ slide(s)

Give your answer ___ slides

5.

2.3 ← ___ slide(s)
× 6.4 ← ___ slide(s)

Give your answer ___ slides

2.

1.6 2 ← ___ slide(s)
× 7.5 ← ___ slide(s)

Give your answer ___ slides

6.

1 3.2 ← ___ slide(s)
× 9 ← ___ slide(s)

Give your answer ___ slides

3.

4.2 ← ___ slide(s)
× .0 9 ← ___ slide(s)

Give your answer ___ slides

7.

1 0.2 ← ___ slide(s)
× . 6 7 ← ___ slide(s)

Give your answer ___ slides

4.

8.5 ← ___ slide(s)
× 7.6 ← ___ slide(s)

Give your answer ___ slides

8.

5.8 ← ___ slide(s)
× 3.3 ← ___ slide(s)

Give your answer ___ slides

Concept Quiz

1. Whenever you have a problem like **7 - .03** you have to make one slight change before you can subtract the numbers. What is that change?

____

____

2. Why do you have to do this when you have this type of decimal subtraction problem?

____

3. Solve: **6 - .07** Answer: [ ]

4. Whenever you have a decimal multiplication problem like .3 x .7 you have to do these three steps:

Step 1: ______________________________________

Step 2: ______________________________________

Step 3: ______________________________________

5. .07 is the same as 7 DIVIDED by ________

6. .07 is the same as 7 MULTIPLIED by $\frac{1}{\square}$

7. When you MULTIPLY decimals you DON'T line up the decimal points first.

TRUE FALSE

8. To go from a decimal to a percent or a percent to a decimal you have to slide the decimal point _____ times.

9. What in the percent sign tells you this?

____

10. If you have a PERCENT and you want a DECIMAL you slide the decimal to the

RIGHT LEFT

Percent - Decimal Conversions

11. 7% = ☐ (decimal)

12. .06 = ☐ % (percent)

13. .003 = ☐ % (percent)

14. .25 = ☐ % (percent)

DM-10 Problems

15. 54 ÷ 10 = ☐

16. .932 × 100 = ☐

Decimal Multiplication

17. .6 × .08 = ☐

18.

$$\begin{array}{r} 2.49 \\ \times\ 5.3 \\ \hline \end{array}$$

☐

Decimal Dividends

Example

A.

KEEP THE QUOTIENT ANSWERS ABOVE THE RIGHT NUMBERS!

$$\begin{array}{r} 4.2 \\ 2\overline{)8.4} \end{array}$$

---CALLED THE ***DIVIDEND!***

IT FLOATS UP TO THE TOP!

Move the decimal straight up

Steps:

#1 Divide normally. (Be sure to line up the quotient answers correctly.)

#2 After you have divided the numbers, just give your answer a decimal point right above the decimal in the dividend.

Solve the following decimal division problems

1. $2\overline{)82.6}$

2. $3\overline{)9.63}$

3. $2\overline{)7.34}$

4. $4\overline{)64.8}$

5. $5\overline{).095}$

6. $2\overline{)1.046}$

Decimal Dividends

Solve the following decimal division problems

These kinds of problems are common when converting FRACTIONS to DECIMALS

1. $\begin{array}{r} 0. \\ 2\overline{)1.0} \end{array}$

2. $5\overline{)3.0}$

3. $4\overline{)1.00}$

4. $8\overline{)1.000}$

5. $4\overline{)3.00}$

6. $3\overline{)1.000}$ $\frac{1}{3}$

Just solve this to three places

7. $3\overline{)2.000}$ $\frac{2}{3}$

Just solve this to three places

8. $8\overline{)7.000}$

Decimal Divisors

Examples

A. $1.2\overline{)2.4} = 12\overline{)24}$ (quotient 2)

Get rid of the decimal in the divisor by sliding the decimal to the right of the 1s place

Slide here too

B. $.3\overline{)36} = 3\overline{)360.}$ (quotient 3)

You may need to add zeros to slide the decimal

(remember the invisible decimal point to the right of the one's place)

DIVISION PROBLEMS ARE JUST LIKE FRACTIONS—YOU CAN MULTIPLY BOTH SIDES BY ANY NUMBER!

In these examples you are just multiplying both sides by 10

Rewrite and solve the following decimal division problems

$\frac{3.3}{1.1}$ (× 10) = $\frac{33}{11}$ (× 10)

SINCE YOU'RE MULTIPLYING BOTH TOP AND BOTTOM BY THE SAME NUMBER, YOU AREN'T CHANGING THE ANSWER AT ALL!

1. $1.1\overline{)3.3}$

the divisor needs one slide

$\overline{)\quad}$

2. $.12\overline{)48}$

the divisor needs two slides

$\overline{)\quad}$

3. $.6\overline{)30}$

$\overline{)\quad}$

4. $.007\overline{)42}$

$\overline{)\quad}$

5. $.2\overline{)8}$

$\overline{)\quad}$

6. $.001\overline{)6}$

$\overline{)\quad}$

7. $.05\overline{)35}$

$\overline{)\quad}$

8. $.03\overline{)60}$

$\overline{)\quad}$

Decimal Divisors

Rewrite and solve the following decimal division problems

Slide both numbers and divide

THE AMOUNT YOU SLIDE DEPENDS ON THIS NUMBER!

1. $1.2 \overline{)48}$

the divisor needs one slide

2. $.08 \overline{)72}$

the divisor needs two slides

3. $.7 \overline{).49}$

4. $.4 \overline{).452}$

5. $.006 \overline{).0078}$

6. $.011 \overline{)88.44}$

7. $.05 \overline{)846.5}$

8. $.008 \overline{)2.72}$

Why Multiplying Decimal Divisors Works

When you slide decimal divisors it's just like multiplying fractions by the same number

Example

A. $1.2\overline{)48}$

BOTTOM NUMBER OF A FRACTION (pointing to 1.2)

TOP NUMBER OF A FRACTION (pointing to 48)

The TOP number is the one that wears the "division hat"

The outside number ("divisor") goes on the BOTTOM of the fraction

$$\frac{48}{1.2} \; \frac{\times 10}{\times 10} = \frac{480}{12}$$

NOW WE NO LONGER HAVE A DECIMAL IN THE FRACTION (DIVISION PROBLEM)!

Rewrite and the following decimal division problems as fractions and get rid of the decimal points by multiplying

	Convert to a fraction	Multiply by 10, 100, 1000, etc. to get rid of the decimal point	Fraction without a decimal point
1. $.5\overline{)3.5}$ (the divisor needs one slide) →	☐ / ☐	× ___ / × ___	= ☐ / ☐
2. $.09\overline{)27}$ (the divisor needs two slides) →	☐ / ☐	× ___ / × ___	= ☐ / ☐
3. $.006\overline{)4.2}$ (the divisor needs three slides) →	☐ / ☐	× ___ / × ___	= ☐ / ☐

.5 is the Same as One Half

Example

.5 is the same as 1/2 (divide by 2)

A. .5 × 6 = 3

.5 IS ONE HALF; WHAT'S HALF OF 6?

.5 IS THE SAME AS 50%!

Multiply the following numbers by .5

1. .5 × 8 = ☐

2. .5 × 16 = ☐

3. .5 × 10 = ☐

4. .5 × 20 = ☐

5. .5 × 18 = ☐

6. .5 × 22 = ☐

7. .5 × 100 = ☐

8. .5 × 220 = ☐

9. .5 × 60 = ☐

10. .5 × 480 = ☐

DECIMAL REVIEW

11. Which is larger: .7 or .08?

☐ Why?

☐

12. Write the number "three and 42 thousandths"

☐

.25 is the Same as One Fourth

Example

.25 is the same as 1/4 (divide by 4)

A. $.25 \times 40 = 10$

.25 IS ONE FOURTH; WHAT'S A FOURTH OF 40?

.25 IS THE SAME AS 25%!

Multiply the following numbers by .25 or .5

1. $.25 \times 8 =$ ☐

2. $.25 \times 20 =$ ☐

3. $.25 \times 12 =$ ☐

4. $.25 \times 24 =$ ☐

5. $.25 \times 40 =$ ☐

6. $.5 \times 20 =$ ☐

7. $.25 \times 100 =$ ☐

8. $.25 \times 88 =$ ☐

9. $.5 \times 16 =$ ☐

10. $.25 \times 16 =$ ☐

11. $.5 \times 24 =$ ☐

12. $.25 \times 24 =$ ☐

13. $.25 \times 28 =$ ☐

14. $.25 \times 20 =$ ☐

15. $.25 \times 64 =$ ☐

16. $.25 \times 144 =$ ☐

Decimal Multiplication Review

1.

$$\begin{array}{r} 7.2 \\ \times\ 5.9 \\ \hline \end{array}$$

← ___ slide(s)
← ___ slide(s)

Give your answer ___ slides

2.

$$\begin{array}{r} 48.1 \\ \times\ 6 \\ \hline \end{array}$$

← ___ slide(s)
← ___ slide(s)

Give your answer ___ slides

3.

$$\begin{array}{r} 57.9 \\ \times\ .43 \\ \hline \end{array}$$

← ___ slide(s)
← ___ slide(s)

Give your answer ___ slides

4.

$$\begin{array}{r} .96 \\ \times\ 8.2 \\ \hline \end{array}$$

← ___ slide(s)
← ___ slide(s)

Give your answer ___ slides

5.

$$\begin{array}{r} 3.83 \\ \times\ 6.4 \\ \hline \end{array}$$

← ___ slide(s)
← ___ slide(s)

Give your answer ___ slides

6.

$$\begin{array}{r} 3.7 \\ \times\ .09 \\ \hline \end{array}$$

← ___ slide(s)
← ___ slide(s)

Give your answer ___ slides

7.

$$\begin{array}{r} 8.04 \\ \times\ .24 \\ \hline \end{array}$$

← ___ slide(s)
← ___ slide(s)

Give your answer ___ slides

8.

$$\begin{array}{r} 4.8 \\ \times\ 7.5 \\ \hline \end{array}$$

← ___ slide(s)
← ___ slide(s)

Give your answer ___ slides

Decimals Test 1

1. Whenever you have a problem like $.2\overline{)\,6\quad}$ you first have to

____

2. Write $.2\overline{)\,6\quad}$ as a fraction ☐

3. Solve: $.07\overline{)\,69\quad}$ Answer: ☐

4. When you multiply by .5 it's the same as ____________________

5. $.5 \times 48 =$ ☐

6. When you multiply by .25 it's the same as ____________________

7. $.25 \times 32 =$ ☐

Decimals Test 2

1. Whenever you have a problem like 2.4 + .78 you first have to

____

2. Solve: $3.75 + 4.7 = \boxed{}$

3. Whenever you have a problem like $.2\overline{)\,6}$ you first have to

____

4. $5\overline{)\,4.395}$

5. Solve:

$$\begin{array}{r} 4.2 \\ \times .37 \\ \hline \end{array}$$

6. .734 is the same as 734 DIVIDED by __________

. . . or 734 MULTIPLIED by $\frac{1}{\boxed{}}$

8. Whenever you multiply decimals you have to line up the decimals first

TRUE FALSE

Supplementary Pages

- DECIMALS
- PERCENTS
- FRACTIONS

A.

DECIMAL | PERCENT | FRACTION

B.

DECIMAL | PERCENT | FRACTION

C.

DECIMAL | PERCENT | FRACTION

D.

DECIMAL | PERCENT | FRACTION

Last Digit Up Trick

A. Use the "Last Digit Up" trick to write the decimal you are given

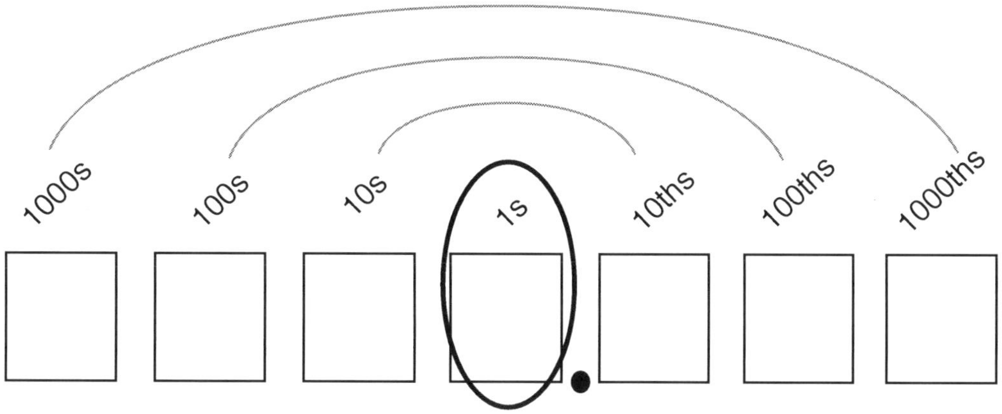

B. Use the "Last Digit Up" trick to write the decimal you are given

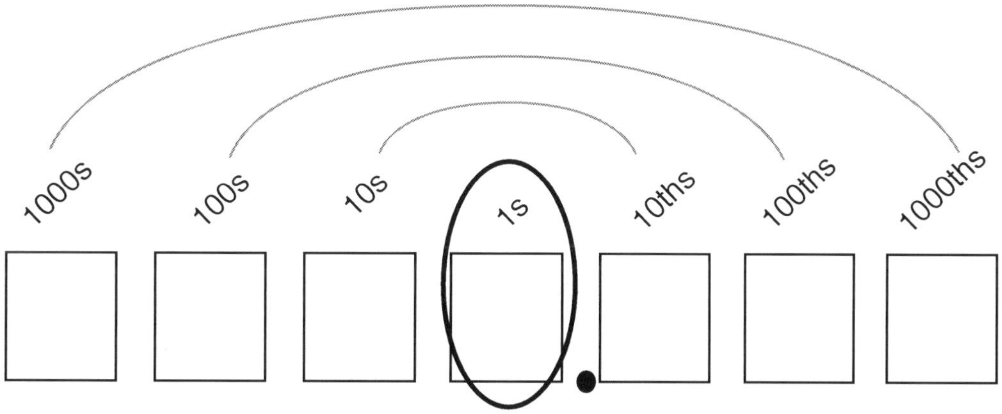

C. Use the "Last Digit Up" trick to write the decimal you are given

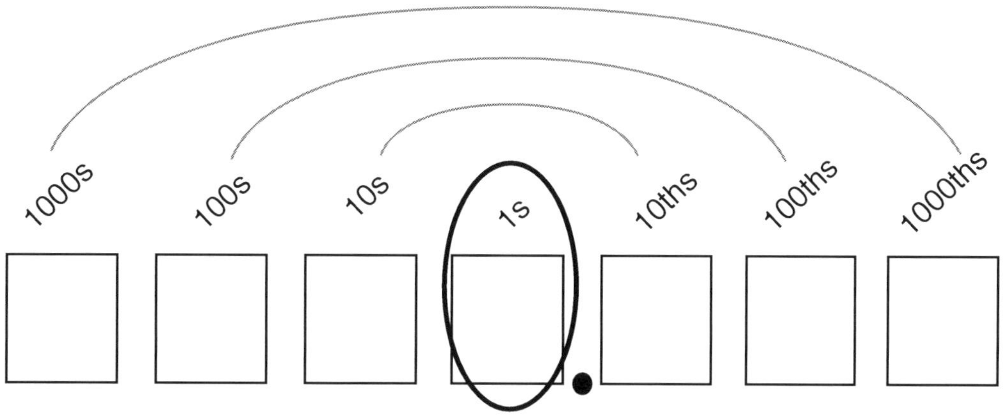

Writing Fractions and Decimals

1.

1000s	100s	10s	1s	.	10ths	100ths	1000ths
				.			

Fraction form: ____________

2.

1000s	100s	10s	1s	.	10ths	100ths	1000ths
				.			

Fraction form: ____________

3.

1000s	100s	10s	1s	.	10ths	100ths	1000ths
				.			

Fraction form: ____________

4.

				.			

Fraction form: ____________

5.

				.			

Fraction form: ____________

Rainbow Chart

Study the following examples of the Rainbow Chart

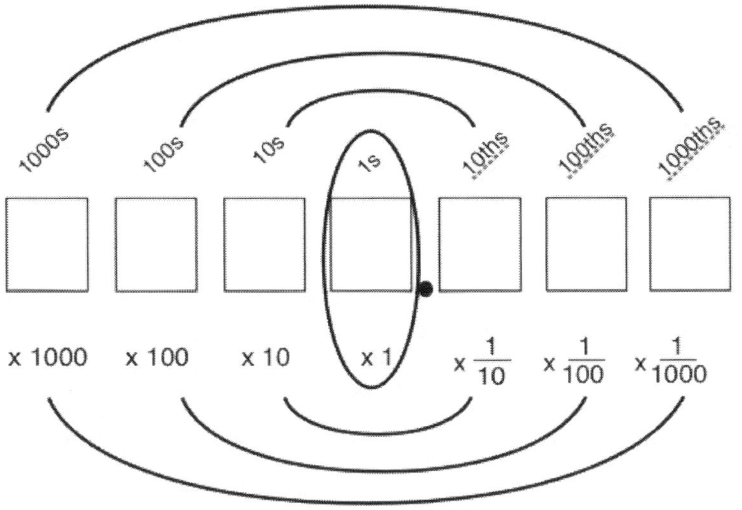

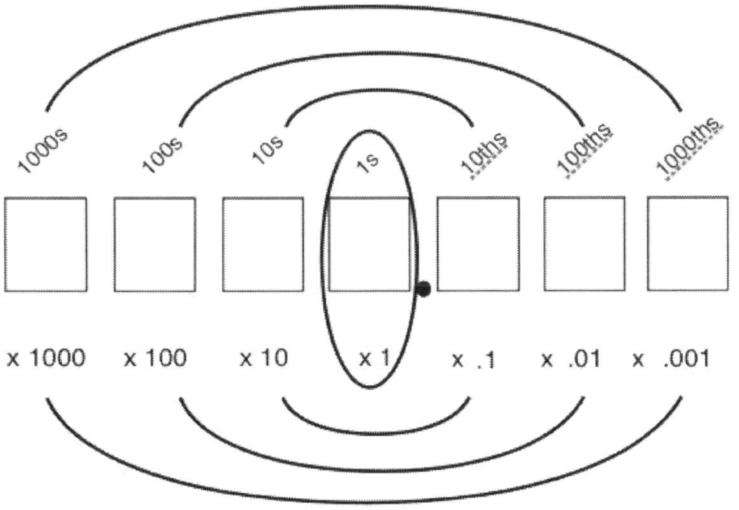

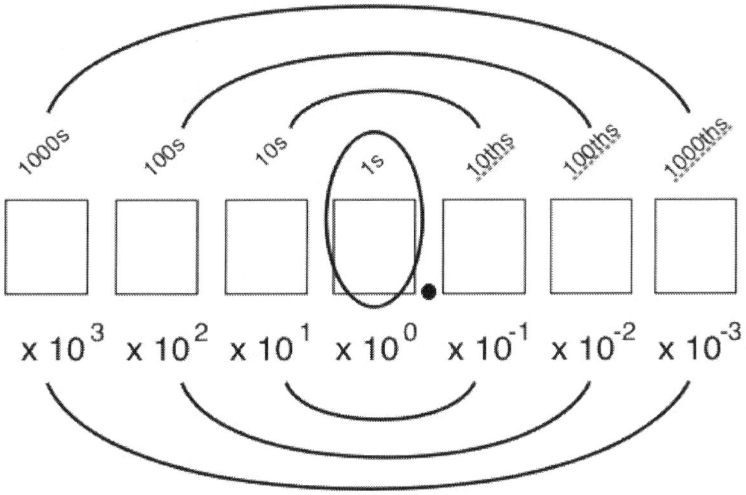

Rainbow Chart

Rainbow Chart Variation #1

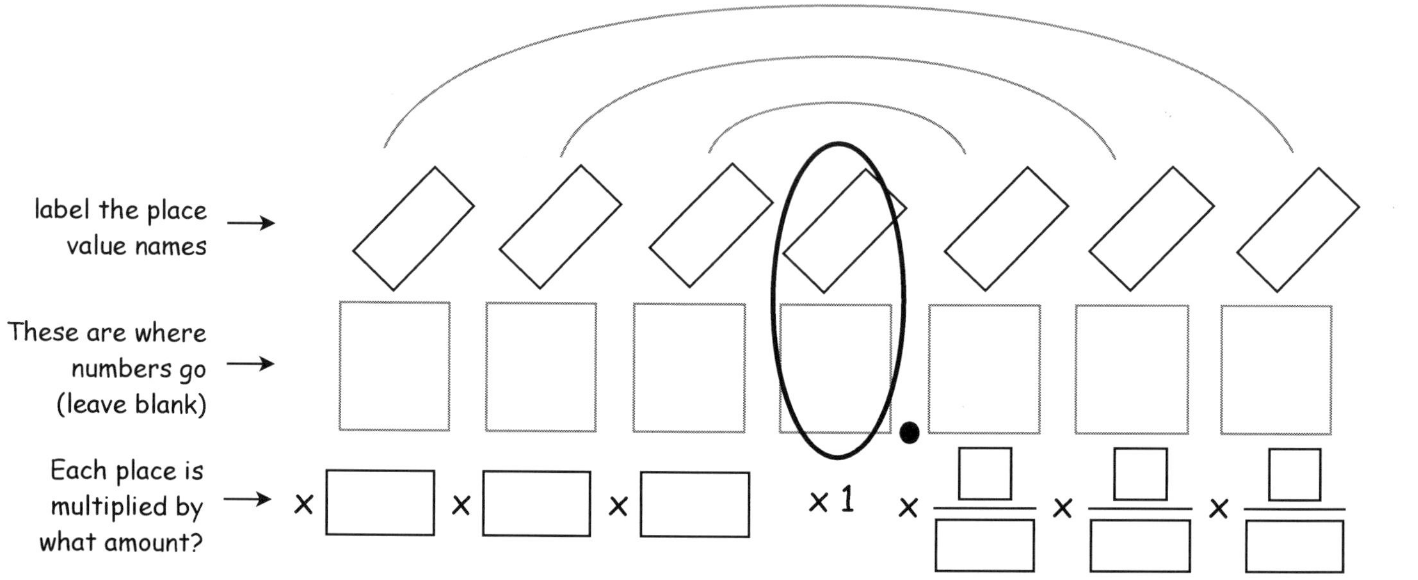

The decimal places are multiplied by what fractions?

Rainbow Chart Variation #2

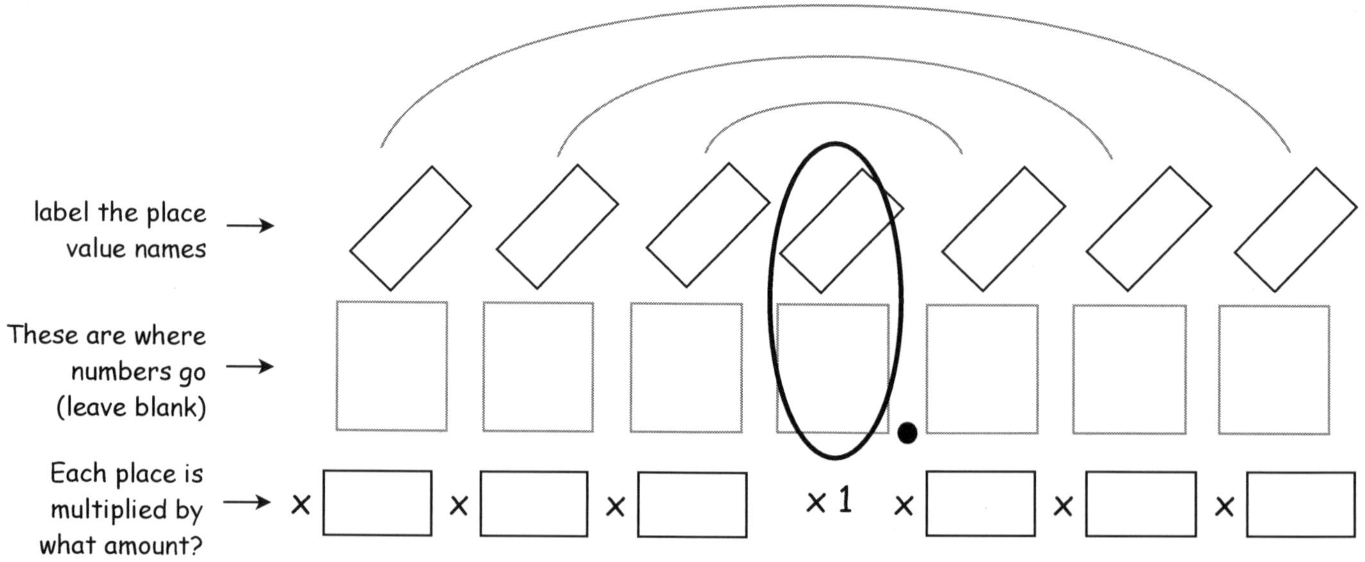

The decimal places are multiplied by what decimals?

Rainbow Chart

Rainbow Chart Variation #3

label the place value names →

These are where numbers go (leave blank) →

Each place is multiplied by what amount? →

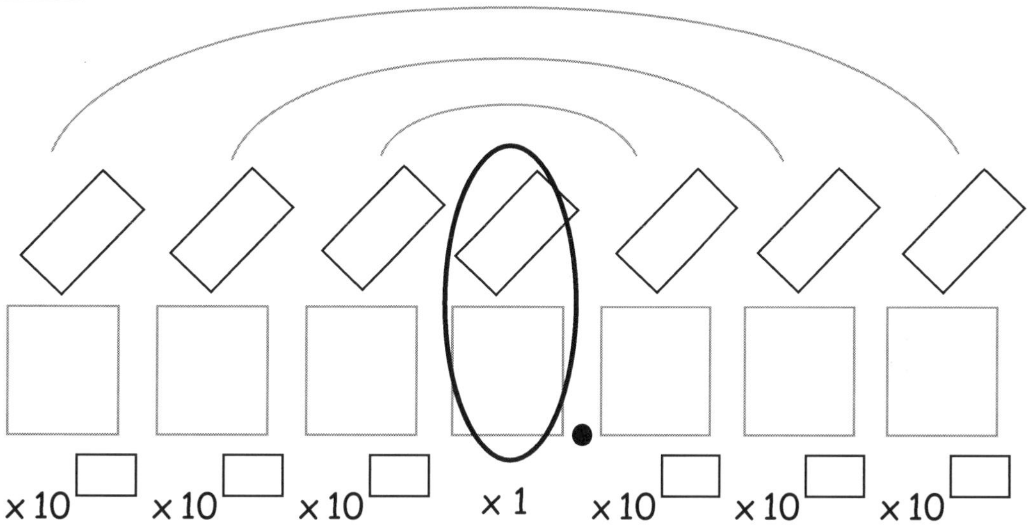

The places are multiplied by which powers of 10?

Rainbow Chart Variation #4 (Base Systems—Advanced)

label the places value names according to the base number you are given →

These are where numbers go (leave blank) →

Each place is multiplied by what amount? →

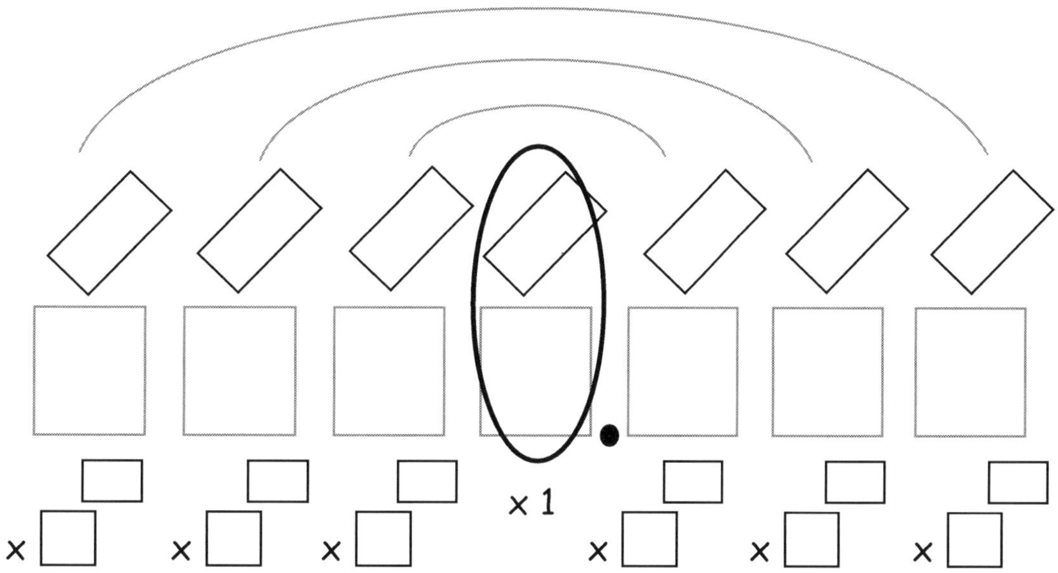

Answer Key

for

MathWise Decimals

What are Decimals Anyway?

• Decimals are just another way of writing fractions (and percents)

A LOT OF MATH INVOLVES LEARNING HOW TO CHANGE COSTUMES!

Numbers wear three costumes

FRACTIONS	DECIMALS	PERCENTS
$2\frac{1}{10}$	2.1	210%

YOU CAN CALL THEM "FDP"

• The same actor (number) can wear any of these 3 costumes.

• He may look different in these costumes—but he's really the same actor!

• Instead of writing $\frac{1}{10}$, you can just write 1 to the RIGHT of the DECIMAL POINT! (.1)

• Decimals are just ways of writing fractions that have 10, 100, 1000, etc. as the denominator. (You'll learn about this later!)

Change from the FRACTION costume to the DECIMAL costume

1. $\frac{3}{10}$ = . **3**
2. $\frac{9}{100}$ = . 0 **9**
3. $\frac{7}{1000}$ = . 0 0 **7**
4. $\frac{23}{100}$ = . **2** **3**

Using Decimal Bars

Models can help to visualize decimals

A. Shade the correct number of bars to show .3

B. Shade the correct number of bars to show 1.2

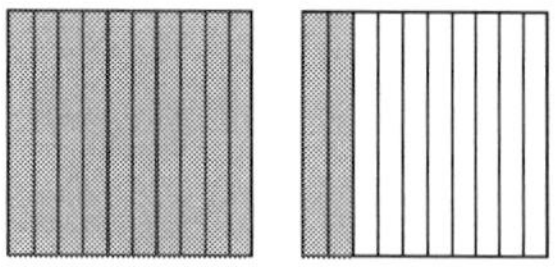

Shade the correct number of bars to show the decimal

1. .4

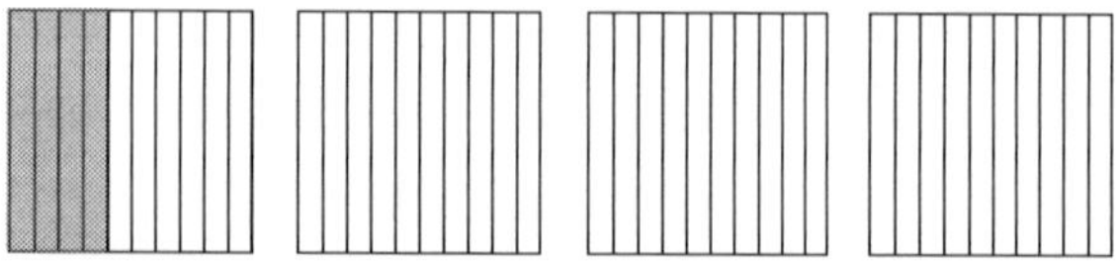

2. 2.8

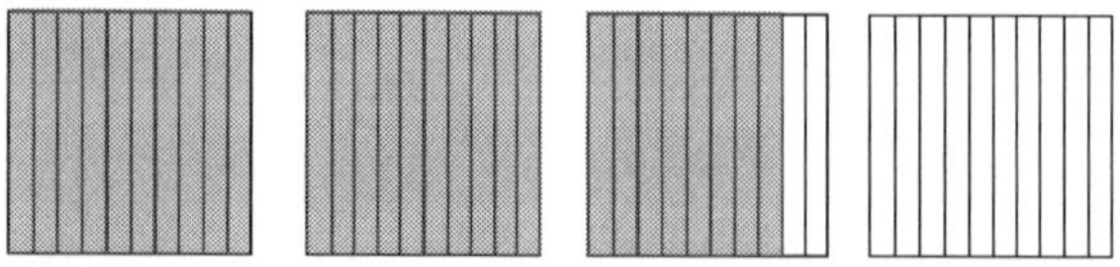

3. 3.7

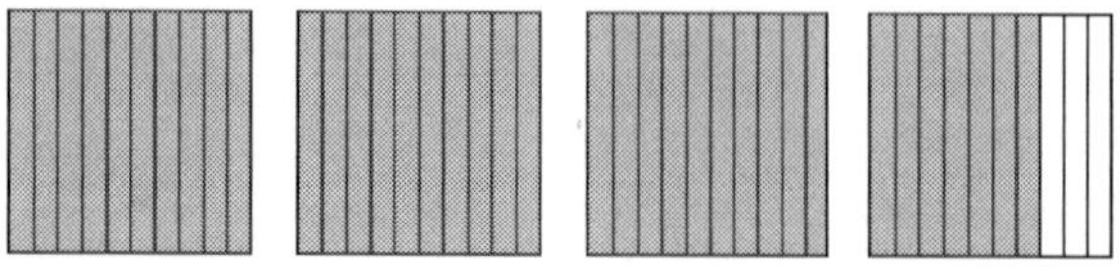

Using Decimal Bars

Write the correct decimal amounts

1. **1.6**

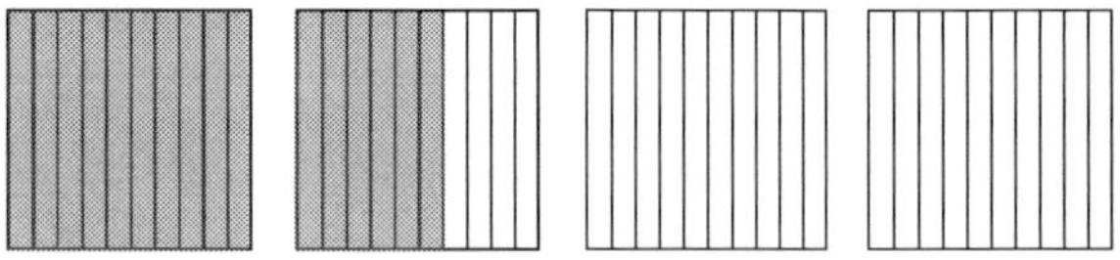

2. **2.9**

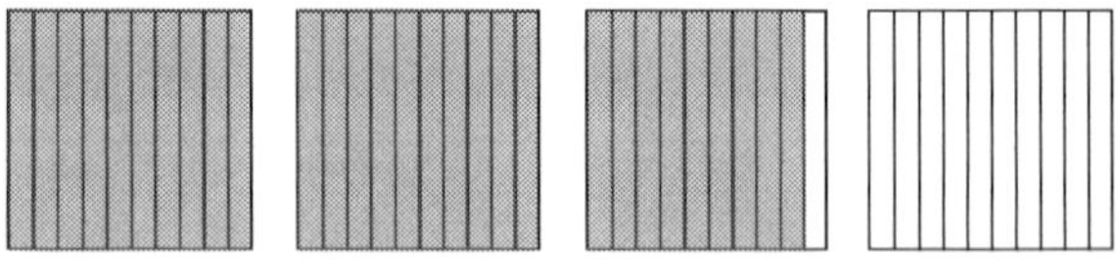

3. **3.4**

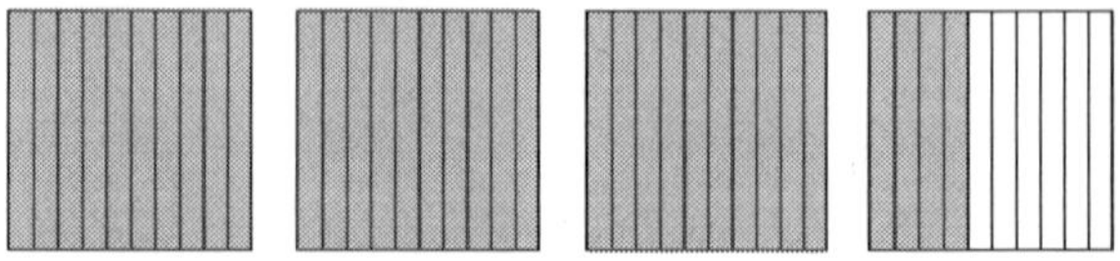

4. **2.5**

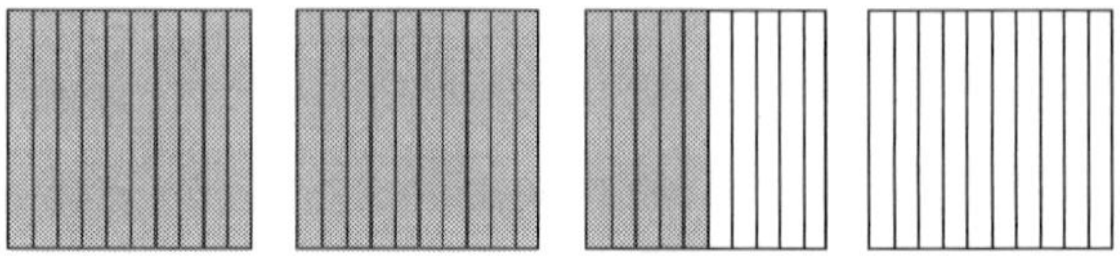

5. **3.8**

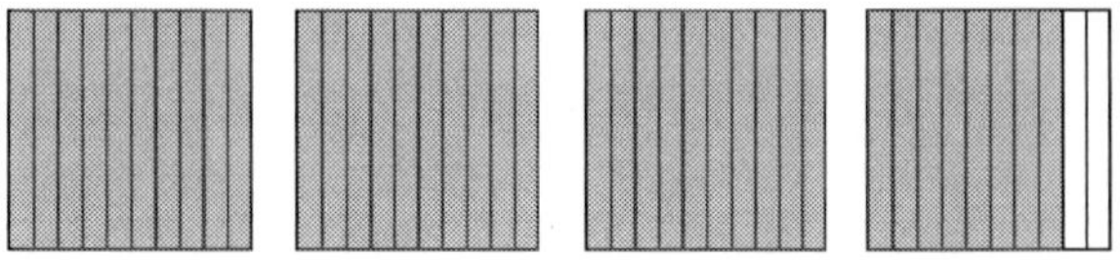

Using Decimal Boxes

A. Shade the correct number of boxes to show .07

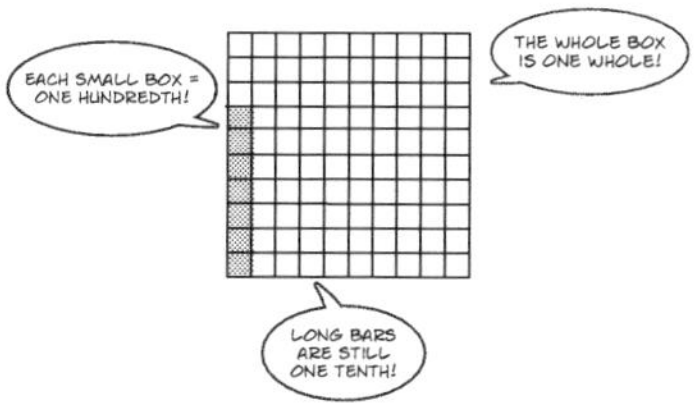

B. Shade the correct number of boxes to show 1.02

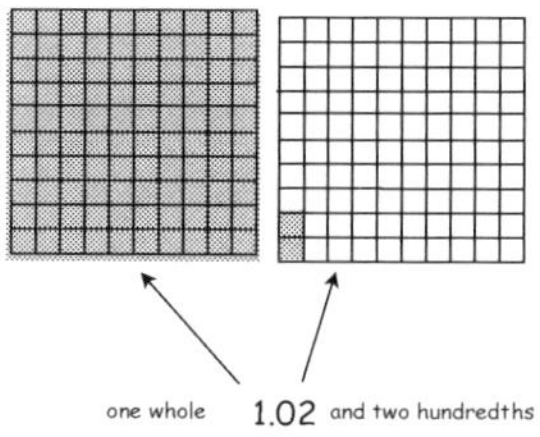

Shade the correct number of boxes to show the decimal

1. .04

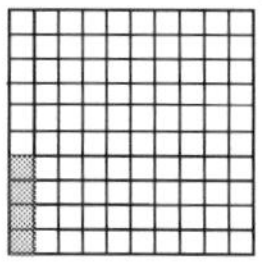

4. .56

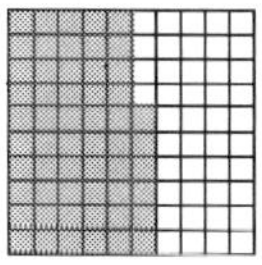

2. .40 (or .4)

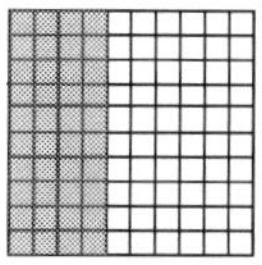

5. .88

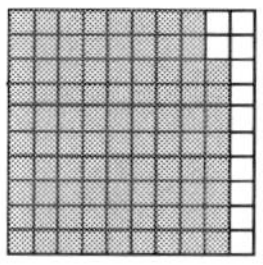

3. 2.75

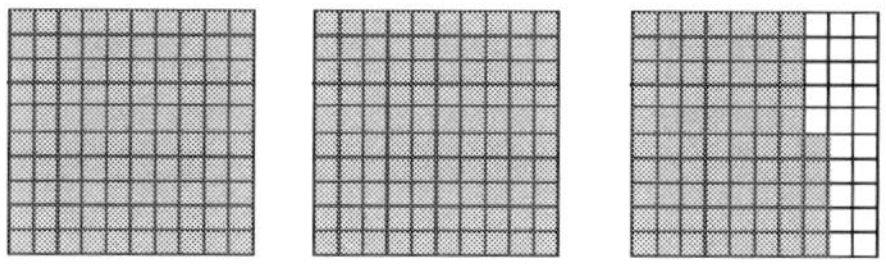

Using Decimal Boxes

Examples

A. .03

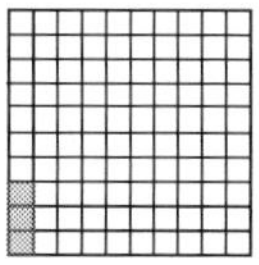

B. .3 (or .30)

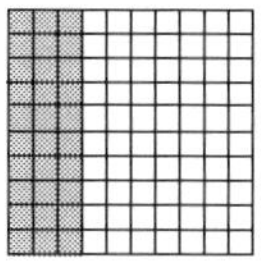

.3 three long bars are the same as...
.30 30 small boxes

Write the correct number for each decimal

1. **.07**

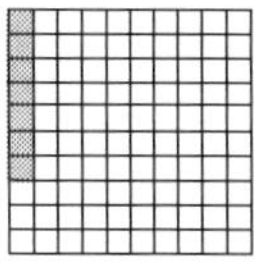

5. **.80** or **.8**

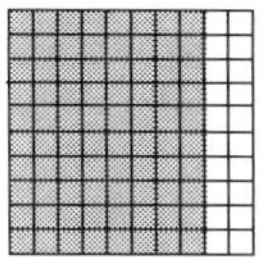

2. **.13**

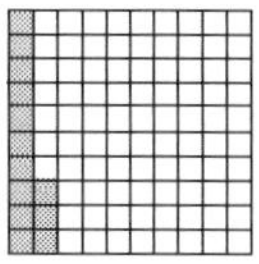

6. **.02**

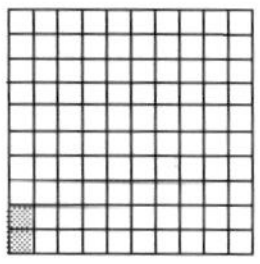

3. **.40** or **.4**

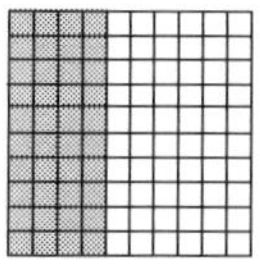

What decimal UNshaded? **.6** or **.60**

7. **.78**

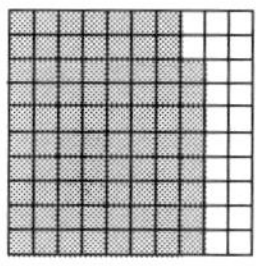

What decimal UNshaded? **.22**

4. **.97**

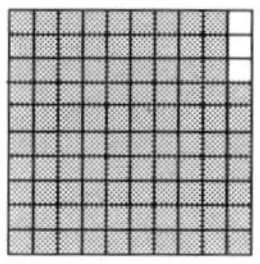

What decimal UNshaded? **.03**

8. **.35**

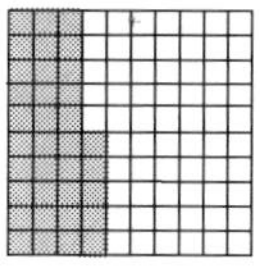

What decimal UNshaded? **.65**

Introduction to the Rainbow Chart

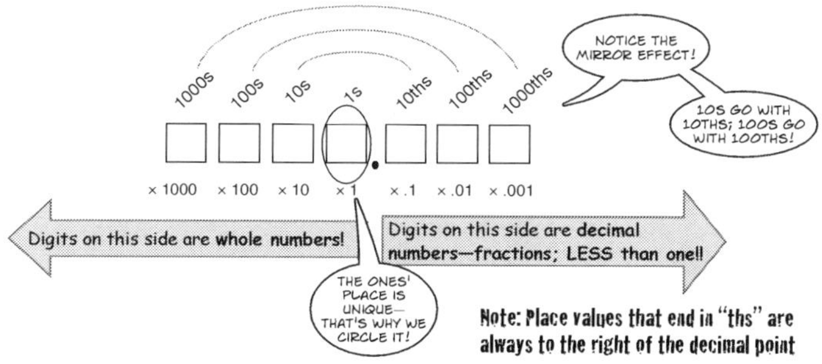

Note: Place values that end in "ths" are always to the right of the decimal point

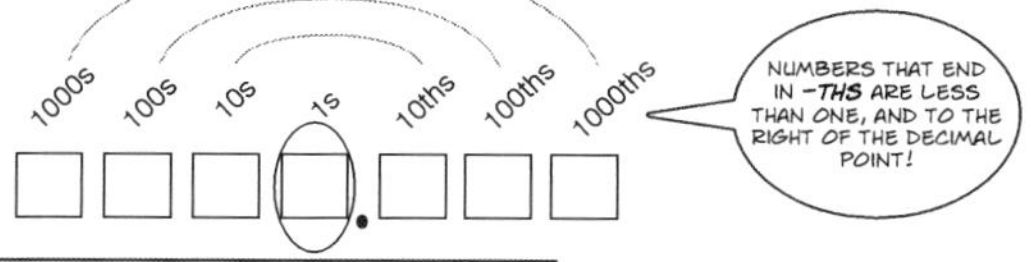

PROBLEM #1: CONNECT THE BOXES WITH RAINBOW LINES. FILL IN THE BOXES AS IN THE EXAMPLES ABOVE.

PROBLEMS 2–3: PUT THE DIGITS IN THE CORRECT PLACES.

1.

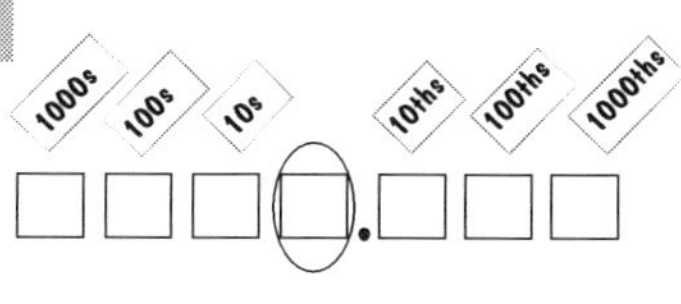

2. Put 3 in the ones' place; 7 in the tens' place; 8 in the tenths' place; 6 in the hundreds' place; 9 in the hundredths' place

3. Put 4 in the thousands place, 7 in the thousandths place, 5 in the hundreds place, 9 in the hundredths place, 8 in the tens place, 2 in the tenths place and 1 in the ones place →

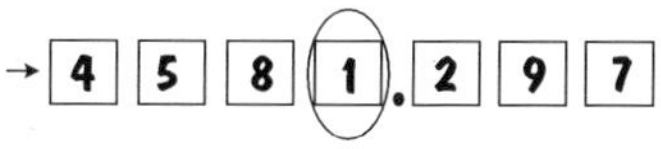

Rainbow Chart

Remember: Place values that end in "ths" are always to the right of the decimal point

1. Label the boxes with the correct place value and draw rainbow lines. Show the multiplication associated with each place value box. Do you see how closely related the names of the rainbow pairs are?

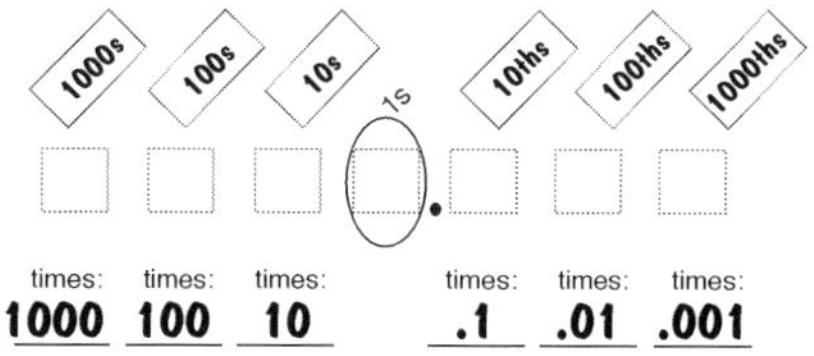

2. Put 6 in the tenths place;
put 2 in the thousandths place.
You can't have a blank in the hundredths place...
so what do you need to put there?

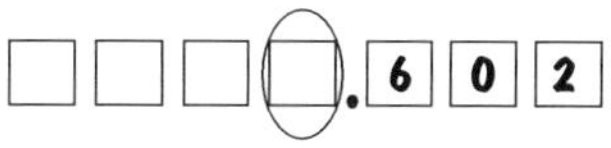

TO WRITE THE NUMBER CORRECTLY YOU NEED TO INCLUDE THREE ZEROS. DO YOU KNOW WHERE THEY GO?

3. Put 7 in the hundreds place;
put 3 in the hundredths place;
put 1 in the thousandths place.

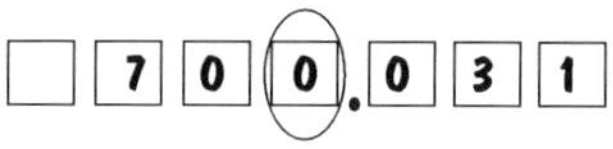

4. Put 6 in the thousandths place;
put 4 in the hundredths place;
put 8 in the tenths place.

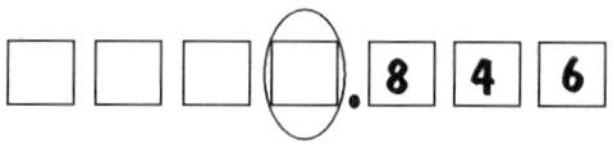

NOW PRACTICE WITH PAGES 75–76
(THOSE PAGES ARE REPRODUCIBLE)

Last Digit Up Trick

A. 27 hundredths

READ THE NUMBER AS JUST A PLAIN "27"— THEN READ THE PLACE NAME ABOVE THE LAST DIGIT (7)!

1000s 100s 10s 1s 10ths 100ths 1000ths

☐ ☐ ☐ ☐ . 2 7 ☐

NOTE: EACH BOX CAN ONLY HAVE 1 DIGIT IN IT!

The last digit of the number points to the correct place

B. 48 thousandths

1000s 100s 10s 1s 10ths 100ths 1000ths

☐ ☐ ☐ ☐ . 0 4 8

You can't have blanks— put a zero here instead

The last digit of the number points to the correct place

Write the decimals in the correct places

to read the decimals, just read the number, then name the place of the last digit

Write the number 98 so that the last digit (8) points to the hundredths place

1. 98 hundredths

1000s 100s 10s 1s 10ths 100ths 1000ths

☐ ☐ ☐ ☐ . **9** **8** ☐

2. 274 thousandths
(Draw rainbow lines below first; then write the number in the correct place)

1000s 100s 10s 1s 10ths 100ths 1000ths

☐ ☐ ☐ ☐ . **2** **7** **4**

3. 4 thousandths

☐ ☐ ☐ ☐ . **0** **0** **4**

4. 45 thousandths

☐ ☐ ☐ ☐ . **0** **4** **5**

5. 6 hundredths

☐ ☐ ☐ ☐ . **0** **6** ☐

6. 208 thousandths

☐ ☐ ☐ ☐ . **2** **0** **8**

Last Digit Up Trick

A. Circle the place of the last digit

1000s 100s 10s 1s 10ths 100ths 1000ths

☐ ☐ ☐ ☐ . 4 8 ☐

48 tenths (hundredths) thousandths

THE LAST DIGIT POINTS UP TO THE THOUSANDTHS PLACE!

B. Circle the place of the last digit

1000s 100s 10s 1s 10ths 100ths 1000ths

☐ ☐ ☐ ☐ . 0 5 6

56 tenths hundredths (thousandths)

Skip the zero (56 is really the same as 056)!

The last digit of the number points to the correct place

Draw rainbow lines & circle the correct place for the LAST DIGIT of each decimal

1. Circle the place of the last digit

1000s 100s 10s 1s 10ths 100ths 1000ths

☐ ☐ ☐ ☐ . 2 3 ☐

23 tenths (hundredths) thousandths

2. Circle the place of the last digit

1000s 100s 10s 1s 10ths 100ths 1000ths

☐ ☐ ☐ ☐ . 8 6 5

865 tenths hundredths (thousandths)

3. Circle the place of the last digit

1000s 100s 10s 1s 10ths 100ths 1000ths

☐ ☐ ☐ ☐ . 0 7 8

78 tenths hundredths (thousandths)

4. Circle the place of the last digit

1000s 100s 10s 1s 10ths 100ths 1000ths

☐ ☐ ☐ ☐ . 3 0 9

309 tenths hundredths (thousandths)

Last Digit Up Trick

Write the number and circle the correct place for the LAST DIGIT of each decimal number

A. Circle the place of the last digit

1000s 100s 10s 1s 10ths 100ths 1000ths

☐ ☐ ☐ ☐ . 0 5 ☐

5 tenths (hundredths) thousandths

1. Draw rainbow lines, write the number, and circle the place of the last digit

1000s 100s 10s 1s 10ths 100ths 1000ths

☐ ☐ ☐ ☐ . 0 4 9

49 tenths hundredths thousandths

2. (1) Draw rainbow lines and label the places
(2) Write the number and circle the correct place

1000 s 100 s 10 s 1s 10 ths 100 ths 1000 ths

☐ ☐ ☐ ☐ . 2 3 ☐

23 tenths hundredths thousandths

3. (1) Draw rainbow lines and label the places
(2) Write the number and circle the correct place

1000 s 100 s 10 s 1s 10 ths 100 ths 1000 ths

☐ ☐ ☐ ☐ . 8 7 ☐

87 tenths hundredths thousandths

4. (1) Draw rainbow lines and label the places
(2) Write the number and circle the correct place

1000 s 100 s 10 s 1s 10 ths 100 ths 1000 ths

☐ ☐ ☐ ☐ . 8 ☐ ☐

8 tenths hundredths thousandths

5. Write the number and circle the place of the last digit

☐ ☐ ☐ ☐ . 5 0 4

504 tenths hundredths thousandths

Concept Quiz

1. Write the decimal amount shaded

2.7

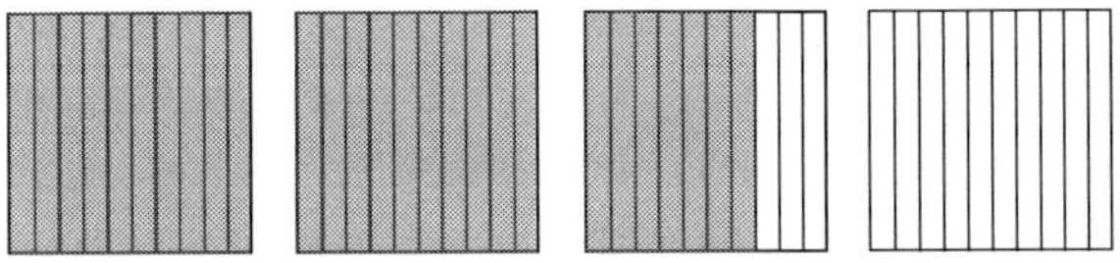

2. 3.6 Shade in the correct amount

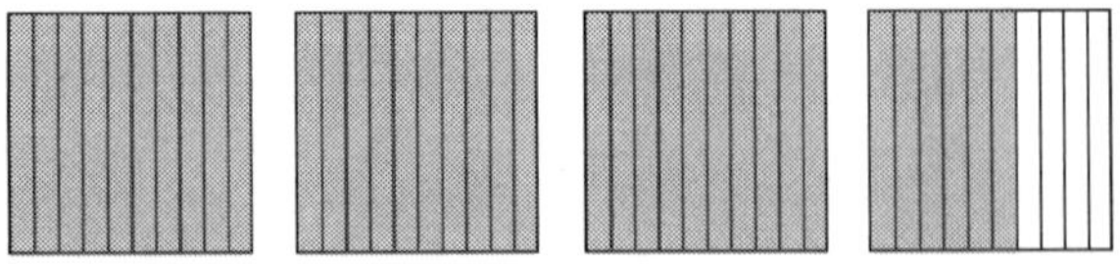

3.

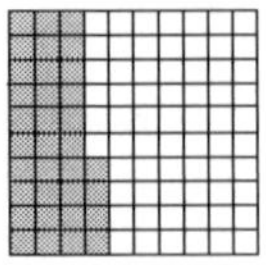

Decimal amount shaded: **.34**

Decimal amount UNshaded: **.66**

4. Label the boxes with the correct place value and draw rainbow lines. Show the multiplication associated with each place value box.

1000s 100s 10s 1s 10ths 100ths 1000ths

☐ ☐ ☐ ☐ . ☐ ☐ ☐

times:	times:	times:		times:	times:	times:
1000	**100**	**10**		**.1**	**.01**	**.001**

Naming Decimals

Put the number in the box and then circle the correct name

REMEMBER: "LAST DIGIT UP"!

USE THIS CHART FOR THE FOLLOWING PROBLEMS!

1000s	100s	10s	1s	.	10ths	100ths	1000ths
				.			

	Decimal	Box			
1.	.5	5	(tenths)	hundredths	thousandths
2.	.04	4	tenths	(hundredths)	thousandths
3.	.27	27	tenths	(hundredths)	thousandths
4.	.3	3	(tenths)	hundredths	thousandths
5.	.685	685	tenths	hundredths	(thousandths)
6.	.81	81	tenths	(hundredths)	thousandths
7.	.008	8	tenths	hundredths	(thousandths)
8.	.046	46	tenths	(hundredths)	thousandths
9.	.9	9	(tenths)	hundredths	thousandths
10.	.937	937	tenths	hundredths	(thousandths)

Writing Decimals

Put the decimal digits in the correct boxes

		1000s	100s	10s	1s	.	10ths	100ths	1000ths
1.	25 hundredths					.	2	5	
2.	25 thousandths					.	0	2	5
3.	379 thousandths					.	3	7	9
4.	17 hundredths					.	1	7	
5.	58 thousandths					.	0	5	8
6.	906 thousandths					.	9	0	6

DON'T FORGET ZERO AS A PLACEHOLDER! —YOU CAN'T HAVE BLANKS!

"And" Means Decimal Point

Examples

LEFT OF THE DECIMAL ← → RIGHT OF THE DECIMAL

A. 4 [and] 7 tenths

WHOLE NUMBER PART ↑ DECIMAL (FRACTION) PART

"AND" MEANS DECIMAL POINT!

[4] . [7]

B. 23 [and] 84 hundredths

whole number — decimal point — decimal part

[2] [3] . [8] [4]

Structure of Whole Number & Decimal Numbers

[whole number] . [decimal part]

↑ "and"

Put the decimal digits in the correct boxes

		1000s	100s	10s	1s	.	10ths	100ths	1000ths
1.	6 and 4 tenths				6	.	4		
2.	9 and 37 thousandths				9	.	0	3	7
3.	89 and 71 hundredths			8	9	.	7	1	

Writing Decimals

Now put digits on both sides of the decimal point

Put the decimal digits in the correct boxes

		1000s	100s	10s	1s	.	10ths	100ths	1000ths
1.	17 and 2 tenths			1	7	.	2		
2.	28 and 13 thousandths			2	8	.	0	1	3
3.	89 and 71 hundredths			8	9	.	7	1	
4.	4 and 506 thousandths				4	.	5	0	6
5.	605 and 27 thousandths		6	0	5	.	0	2	7
6.	76 and 49 hundredths			7	6	.	4	9	

Concept Quiz

1. Put the number in the box and then circle the correct name

.237 **237** tenths hundredths (thousandths)

2. Put the digits in the correct boxes

1000s 100s 10s 1s 10ths 100ths 1000ths

48 hundredths → [ ] [ ] [ ] ([ ]) . [**4**] [**8**] [ ]

3. [ ] [ ] [ ] ([ ]) . [ ] [ ] [ ]

Every number in this box is really multiplied by: **.001**

4. Write the number "9 and 2 tenths" [ ] [ ] [ ] [**9**] . [**2**] [ ] [ ]

5. Write the number "7 and 5 hundredths" [ ] [ ] [ ] [**7**] . [**0**] [**5**] [ ]

6. Write the number "34 and 207 thousandths" [ ] [ ] [**3**] [**4**] . [**2**] [**0**] [**7**]

The DM-10 Trick

D – DIVIDE
M – MULTIPLY
10 – WITH TEN OR POWERS OF TEN (100'S, 1000S, 10THS, 100THS, ETC.)

IF YOU MULTIPLY OR DIVIDE A NUMBER BY A POWER OF 10, YOU JUST SLIDE THE DECIMAL ON THE ORIGINAL NUMBER!

D (DIVIDE) ←	M (MULTIPLY) →
When you divide by 10s you slide to the LEFT	When you multiply by 10s you slide to the RIGHT

THIS IS WHY THE "D" COMES FIRST IN "DM-10"!

LET'S START BY DIVIDING BY 10S

WHEN YOU DIVIDE, YOU SLIDE TO THE LEFT!

Examples

A. 5.92 ÷ 10 = .592

ONE ZERO, ONE SLIDE!

B. .98 ÷ 10 = .098

IF YOU NEED TO, ADD ONE OR MORE ZEROS TO GIVE ROOM TO SLIDE

(YOU CAN'T SLIDE IN BLANK SPACE!)

1. 62 ÷ 10 = **6.2**

REMEMBER! THERE IS AN INVISIBLE DECIMAL POINT HERE!

2. 4 ÷ 10 = **.4**

3. .7 ÷ 10 = **.07**

4. 700 ÷ 10 = **70**

5. 700 ÷ 100 = **7**

TWO ZEROS, TWO SLIDES!

6. 84.5 ÷ 100 = **.845**

7. 4 ÷ 100 = **.04**

8. 937 ÷ 1,000 = **.937**

STILL TO THE LEFT BECAUSE WE ARE DIVIDING!

HOW MANY SLIDES DO YOU THINK WITH THREE ZEROS?

9. 214.8 ÷ 1,000 = **.2148**

TIP: IT HELPS TO DRAW ARROWS FOR EACH SLIDE!

The DM-10 Trick with Multiplication

D (DIVIDE) ←	M (MULTIPLY) →
When you divide by 10s you slide to the LEFT	When you multiply by 10s you slide to the RIGHT

LET'S START BY MULTIPLYING BY 10S

WHEN YOU MULTIPLY, YOU SLIDE TO THE RIGHT!

Examples

A. 2.46 × 10 = 24.6

ONE ZERO, ONE SLIDE!

B. 15 × 10 = 150

other ways of writing 15: 15. 15.0

1. .65 · 10 = **6.5**

2. .872 · 100 = **87.2**

3. .91 · 100 = **91**

4. .2 · 100 = **20**

5. 30 · 10 = **300**

6. .07 · 10 = **.7**

7. .01 · 100 = **1**

8. .0053 · 100 = **.53**

9. .04 · 1,000 = **40**

10. 7.52 · 1,000 = **7,520**

11. 29.2 · 100 = **2,920**

12. 813.4 · 100 = **81,340**

The DM-10 Trick with Multiplication & Division

D (DIVIDE) ←	M (MULTIPLY) →
When you divide by 10s you slide to the LEFT	When you multiply by 10s you slide to the RIGHT

1. 2.87 · 10 = **28.7**

2. 36.2 ÷ 10 = **3.62**

3. 365 ÷ 100 = **3.65**

4. .86 ÷ 10 = **.086**

5. 4.72 · 10 = **47.2**

6. 3.5 · 100 = **350**

7. 3.5 ÷ 100 = **.035**

8. .009 · 100 = **.9**

9. .1 · 10 = **1**

10. .003 · 10 = **.03**

11. .003 · 100 = **.3**

12. .01 · 100 = **1**

13. 7 ÷ 100 = **.07**

14. .17 · 1000 = **170**

15. 781 ÷ 10 = **78.1**

16. .047 · 100 = **4.7**

17. 8.94 · 10 = **89.4**

18. .2 ÷ 100 = **.002**

Seeing Decimals as Division

Place Value

SHOW THE DIVISION PROBLEM INDICATED BY EACH DECIMAL

CIRCLE THE CORRECT ANSWER!

1.	.3 = 3 ÷	(10)	100	1000
2.	.1 = 1 ÷	(10)	100	1000
3.	.01 = 1 ÷	10	(100)	1000
4.	.52 = 52 ÷	10	(100)	1000
5.	.347 = 347 ÷	10	100	(1000)
6.	.092 = 92 ÷	10	100	(1000)
7.	.25 = 25 ÷	10	(100)	1000
8.	.602 = 602 ÷	10	100	(1000)
9.	.06 = 6 ÷	10	(100)	1000
10.	.007 = 7 ÷	10	100	(1000)
11.	.4 = 4 ÷	(10)	100	1000

WRITE THE DECIMALS:

12.	4 ÷ 10	**.4**
13.	4 ÷ 100	**.04**
14.	27 ÷ 10	**2.7**
15.	27 ÷ 100	**.27**
16.	419 ÷ 1000	**.419**
17.	6 ÷ 100	**.06**
18.	12 ÷ 100	**.12**
19.	5 ÷ 1000	**.005**
20.	19 ÷ 1000	**.019**
21.	2 ÷ 10	**.2**
22.	425 ÷ 1000	**.425**

Decimals as Dividing by Multiples of Ten

WHEN YOU DIVIDE, YOU SLIDE TO THE LEFT! (ONE SLIDE FOR EACH ZERO!)

THIS IS SIMILAR TO THE "DM-10 TRICK"!

A. .6 = 6 ÷ 10 or $6 \cdot \frac{1}{10}$

THE BOTTOM NUMBERS OF FRACTIONS HAVE THE IDEA OF DIVISION!

Write each decimal as a division and fraction problem

1. .03 = 3 ÷ **100** or $3 \cdot \frac{1}{100}$

2. .42 = 42 ÷ **100** or $42 \cdot \frac{1}{100}$

3. .002 = 2 ÷ **1000** or $2 \cdot \frac{1}{1000}$

4. .039 = 39 ÷ **1000** or $39 \cdot \frac{1}{1000}$

5. .364 = 364 ÷ **1000** or $364 \cdot \frac{1}{1000}$

6. .704 = **704** ÷ **1000** or **704** $\cdot \frac{1}{1000}$

Insights into Decimals

The number of decimal slides match the zeros on the bottom of a fraction

Examples

A. .1 = $\frac{1}{10}$ — one slide... one zero

.3 = $\frac{3}{10}$ — one slide... one zero

B. .01 = $\frac{1}{100}$ — two slides... two zeros

.07 = $\frac{7}{100}$ — two slides... two zeros

C. .001 = $\frac{1}{1000}$ — three slides... three zeros

.004 = $\frac{4}{1000}$ — three slides... three zeros

NOW YOUR TURN TO PRACTICE:

1. Write .06 as a fraction $\frac{6}{100}$

2. Write $\frac{9}{1000}$ as a decimal **.009**

D. **Example**

1000s, 100s, 10s, 1s, 10ths, 100ths, 1000ths

× 1000, × 100, × 10, × 1, $\times \frac{1}{10}$, $\times \frac{1}{100}$, $\times \frac{1}{1000}$

Same as what we had before: × .1, × .01, × .001

Decimal is 1 slide to the left of the digit in this place

Decimal is 2 slides to the left of the digit in this place

Decimal is 3 slides to the left of the digit in this place

THIS IS ANOTHER WAY TO SHOW THE PLACE VALUE OF THE DECIMAL NUMBERS!

The DZ ("Dizzy") Trick

How to turn decimals into fractions

THE "D-Z TRICK" PUT A *ZERO* UNDER EVERY *DECIMAL DIGIT!*

THEN CHANGE THE DECIMAL POINT TO A NUMBER ONE

AND ADD A FRACTION LINE! YOU'RE DONE!

IT WORKS SO FAST, IT MAY MAKE YOU FEEL D-Z ("DIZZY")!

THREE NUMBERS TO THE RIGHT OF THE DECIMAL!

NUMBERS TO THE *LEFT* OF THE DECIMAL POINT ARE *WHOLE NUMBERS*--BIG NUMBERS BEFORE THE FRACTION!

A. .583 → 1000 (TURN THE DECIMAL INTO A 1!) (WRITE A ZERO UNDER EACH DECIMAL NUMBER!)

$= \frac{583}{1000}$

B. 2.918 = $2\frac{918}{1000}$ (000 under 918)

- **#1** Put a zero under every decimal digit
- **#2** Trade the decimal point for a 1 to the LEFT of the zeros
- **#3** Draw a fraction bar

Write the following decimals as fractions

1. .467 = $\frac{467}{1000}$

2. .742 = $\frac{742}{1000}$

3. .61 = $\frac{61}{100}$

4. 3.792 = $3\frac{792}{1000}$

5. 65.83 = $65\frac{83}{100}$

6. 87.569 = $87\frac{569}{1000}$

7. 7.2836 = $7\frac{2836}{10000}$

The DP Trick

DR. DECIMAL

2 slides to the **LEFT**, if you have a PERCENT and want to make a DECIMAL

It takes 2 slides ⟷ ⟷ to get from one side to another!

PEPPER PERCENT

2 slides to the **RIGHT**, if you have a DECIMAL and want to make a PERCENT

Use the DP Trick to change percents to decimals and the other way around

	DR. DECIMAL		PEPPER PERCENT
A.	.60 (or .6)	=	60% (60.%)

slide the decimal point 2 times to the LEFT

	DR. DECIMAL		PEPPER PERCENT
1.	**.4**	slide the decimal point 2 times to the LEFT	40%
2.	**.35**	slide the decimal point 2 times to the LEFT	35%
3.	**.07**	slide the decimal point 2 times to the LEFT	7% (07.%)
4.	**4.67**	slide the decimal point 2 times to the LEFT	467%
5.	.20	slide the decimal point 2 times to the RIGHT	**20** %
6.	.84	slide the decimal point 2 times to the RIGHT	**84** %
7.	.3 (= .30)	slide the decimal point 2 times to the RIGHT	**30** %
8.	.05	slide the decimal point 2 times to the RIGHT	**5** %
9.	6.31	slide the decimal point 2 times to the RIGHT	**631** %

THE INVISIBLE DECIMAL POINT STARTS HERE!

YOU NEED TO ADD A ZERO BEFORE THE 6 TO HAVE ROOM TO SLIDE TWO TIMES!

The DP Trick

Just slide two times to go from decimals to percents (or percents to decimals)

Use the DP Trick to change percents to decimals or the other way around

	DR. DECIMAL		PEPPER PERCENT
1.	.48	=	**48** %
2.	**.10**	=	10%
3.	**.38**	=	38%
4.	.05	=	**5** %
5.	.5	=	**50** %
6.	5	=	**500** %
7.	**9.71**	=	971%
8.	**.08**	=	8%
9.	.049	=	**4.9** %
10.	3.2	=	**320** %
11.	**.007**	=	.7%
12.	**1.25**	=	125%
13.	.003	=	**.3** %
14.	15	=	**1500** %
15.	**.186**	=	18.6%
16.	**.637**	=	63.7%
17.	4.19	=	**419** %
18.	**4**	=	400%

Decimals on a Number Line

A.
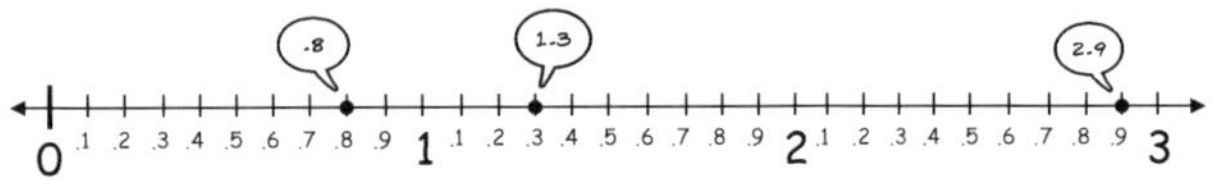

Write the decimal numbers for each of the points on the line

1.
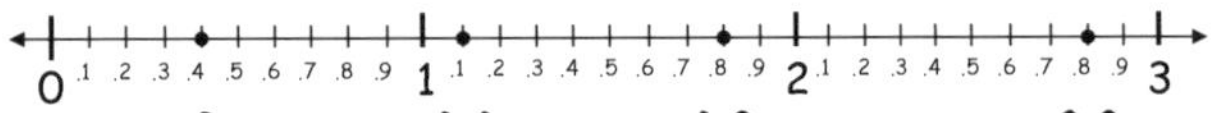

.4 **1.1** **1.8** **2.8**

2.
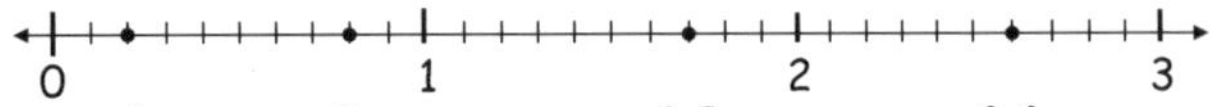

.2 **.8** **1.7** **2.6**

DRAW and LABEL points on the number line for the following decimals

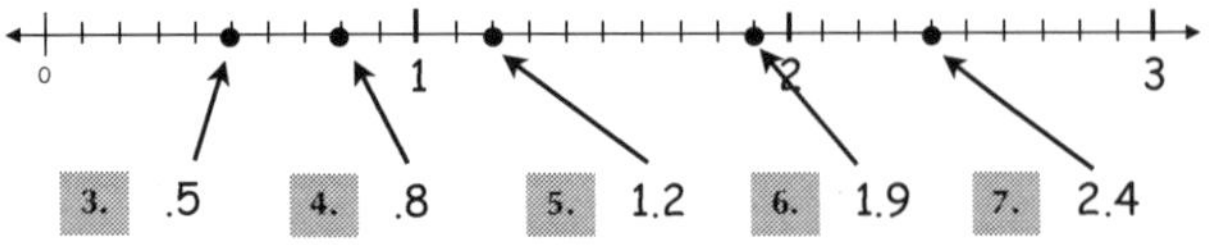

3. .5 4. .8 5. 1.2 6. 1.9 7. 2.4

Continue counting by TENTHS

8. 1.1 1.2 **1.3** 1.4 **1.5** **1.6** **1.7** **1.8** **1.9** **2** **2.1**

Number Triangles with Decimals

Example

A.
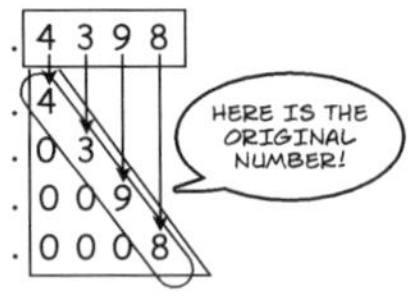

Number triangles help to show expanded notation. They enable you to see how decimal numbers are constructed!

Make number triangles like the example above

1. .8 3 4 2

.**8**
+ .**0** **3**
+ .**0** **0** **4**
+ .**0** **0** **0** **2**

2. .4 5 7 9

.**4**
+ .**0** **5**
+ .**0** **0** **7**
+ .**0** **0** **0** **9**

DO THESE THE EXACT SAME WAY. IGNORE THE BOXES THAT YOU DON'T NEED!

3. .6 3 1 5

.**6**
+ .**0** **3**
+ .**0** **0** **1**
+ .**0** **0** **0** **5**

4. .3 3 0 8

.**3**
+ .**0** **3**
+ .**0** **0** **0**
+ .**0** **0** **0** **8**

Expanded Notation to Standard Notation

Examples

In order (greatest to least)

A. .7 + .02 + .005 + .0006 = .7256

.7
+ .02
+ .005
+ .0006
= .7256

"Expanded Notation" "Standard Notation"

Not in order

B. .004 + .9 + .0001 + .03 = .9341

.9
+ .03
+ .004
+ .0001
.9341

Convert the numbers from EXPANDED NOTATION to STANDARD NOTATION

1. .008 + .5 + .0004 + .03

.5
+ .03
+ .008
+ .0004
= **.5384**

PUT THE DECIMALS IN THE RIGHT PLACES!

2. .0007 + .01 + .6 + .009

.6
+ .01
+ .009
+ .0007
= **.6197**

3. .006 + .7 + .0002

= **.7062**

4. .03 + .008 + .0009 + .2

= **.2389**

Decimals Practice

1. .65 · 10 = **6.5**
(DM-10 Trick)

2. .31 · 1000 = **310**
(DM-10 Trick)

3. 214 ÷ 100 = **2.14**
(DM-10 Trick)

4. 8.7 ÷ 100 = **.087**
(DM-10 Trick)

5. .29 = $\frac{29}{100}$
DZ Trick)

6. .071 = $\frac{71}{1000}$
DZ Trick)

7. .3 = $\frac{3}{10}$
DZ Trick)

8. 4.897 = **4** $\frac{897}{1000}$
DZ Trick)

9. .3 = 3 ÷ **10**

10. .75 = 75 ÷ **100**

11. .45 = **45** %

12. **.07** = 7 %

13. .063 = **6.3** %

14. **.028** = 2.8 %

15. .94 = 94 · $\frac{1}{100}$

16. .127 = 127 · $\frac{1}{1000}$

17. Write this decimal in expanded notation using the Number Triangle Trick

.5193

.5
+ .01
+ .009
+ .0003

Decimals Practice

1. .5 = $\frac{5}{10}$

2. $\frac{7}{10}$ = **.7**

3. .43 = $\frac{43}{100}$

NOTICE THAT THE NUMBER OF ZEROS MATCHES THE NUMBER OF DECIMAL SLIDES!

4. .675 = $\frac{675}{1000}$

5. $\frac{82}{100}$ = **.82**

6. .917 = $\frac{917}{1000}$

7. .028 = $\frac{28}{1000}$

8. $\frac{48}{100}$ = **.48**

9. $\frac{37}{1000}$ = **.037**

10. .917 = $\frac{917}{1000}$

11. Put into standard notation (regular decimal number)

.3 + .01 + .002 + .0005

.3125

12. Put into standard notation (regular decimal number)

.009 + .5 + .0006 + .08

.5896

13. Write the following decimal in expanded notation

.6472

.6 + .04 + .007 + .0006

14. Write 1.618 as a percent:

161.8 %

Adding and Subtracting Elevator

Examples

A. 2.4 + .78

2.4
+ .78
3.18

B. 7.8 − .46

BLANK SPACE MINUS 6??

7.8
− .46

YOU CAN'T HAVE BLANKS WHEN YOU SUBTRACT!

ADD A ZERO SO YOU CAN REGROUP AND THEN SUBTRACT!

7.80
− .46
7.34

WHENEVER YOU ADD OR SUBTRACT DECIMALS YOU NEED A DECIMAL ELEVATOR!

PUT THE DIGITS IN THE CORRECT BOXES!

NOT ALL BOXES WILL BE USED!

ADD ZEROS WHERE THEY BELONG!

1. 5.6 + .37

5.60
+ 0.37
5.97

WRITE ZEROS WHENEVER YOU HAVE BLANKS, ESPECIALLY WHENEVER YOU SUBTRACT DECIMALS!

2. 2.7 − .13

2.70
− 0.13
2.57

3. 3.66 + 5.4

3.66
+ 5.40
9.06

4. 8.45 − 7.6

8.45
− 7.60
0.85

5. 5.4 − 3.44

5.40
− 3.44
1.96

6. 5 − 2.71

5.00
− 2.71
2.29

7. 7.03 − 4.9

7.03
− 4.90
2.13

Adding and Subtracting Elevator

Line up the decimals and add or subtract; put zeros in blank boxes

You line up the decimals when you ADD or SUBTRACT; but NOT when you divide or multiply

1. .38 + 5.7

+	0	.	3	8
	5	.	7	0
	6	.	0	8

PUT ADDITION OR SUBTRACTION SIGNS HERE!

4. 3.41 + 2.071

+	3	.	4	1	0
	2	.	0	7	1
	5	.	4	8	1

2. 7.53 - .44

-	7	.	5	3
	0	.	4	4
	7	.	0	9

5. 8.34 - 4.052

-	8	.	3	4	0
	4	.	0	5	2
	4	.	2	9	0

DON'T FORGET TO ADD A ZERO! YOU CAN'T SUBTRACT WITH BLANKS!

3. 5.07 - .18

-	5	.	0	7
	0	.	1	8
	4	.	8	9

6. 5.6 - 2.037

-	5	.	6	0	0
	2	.	0	3	7
	3	.	5	6	3

Decimal Addition & Subtraction Practice

Line up the decimals, and add or subtract

Line up the numbers correctly before adding or subtracting

1. .24 + 1.9

+ 0 . 2 4
1 . 9 0
Answer: 2 . 1 4

4. 8.67 + .0894

+ 8 . 6 7 0 0
0 . 0 8 9 4
Answer: 8 . 7 5 9 4

2. 8.07 - .65

- 8 . 0 7
0 . 6 5
Answer: 7 . 4 2

5. 9.02 - 3.047

- 9 . 0 2 0
3 . 0 4 7
Answer: 5 . 9 7 3

3. 3.24 - .37

- 3 . 2 4
0 . 3 7
Answer: 2 . 8 7

6. 8.6 - 5.431

- 8 . 6 0 0
5 . 4 3 1
Answer: 3 . 1 6 9

Adding Decimals Mentally

Line up the decimals MENTALLY; then add
Afterwards write the numbers in the boxes to check your answer

1. .3 + .5 = .8

	.	3	
	.	5	
	.	8	

CHECK YOUR ANSWER IN THE BOXES AFTER DOING THE PROBLEM MENTALLY!

4. 0.71 + 0.6 = 1.31

	.	7	1
	.	6	0
1	.	3	1

2. .7 + .7 = 1.4

	.	7	
	.	7	
1	.	4	

5. 1.6 + 0.35 = 1.95

1	.	6	
0	.	3	5
1	.	9	5

3. .06 + .17 = .23

	.	0	6
	.	1	7
	.	2	3

6. 2.64 + 2.6 = 5.24

2	.	6	4
2	.	6	0
5	.	2	4

Adding Decimals Mentally

Line up the decimals MENTALLY; then add

THIS MAY HELP YOU TO VISUALIZE HOW TO LINE UP THE DECIMALS THAT YOU SOLVE MENTALLY ON THIS PAGE!

	.		
	.		
	.		

1. .8 + .08 = .88

2. 3.8 + .8 = 4.6

3. .7 + .03 = .73

4. .7 + .3 = 1.0 or 1

5. .6 + .25 = .85

6. .08 + .06 = .14

7. .18 + .06 = .24

8. .62 + .08 = .70 or .7

8. .02 + .08 = .10 or .1

9. 6.85 + 0.5 = 7.35

10. .03 + .17 = .20 or .2

11. .45 + .45 = .90 or .9

12. 3.6 + 3.6 = 7.2

13. 2.7 + 2.76 = 5.46

14. 49.8 + 0.5 = 50.3

15. 8.96 + .04 = 9.00 or 9

16. 7.35 + .36 = 7.71

17. 5.73 + 1.09 = 6.82

18. 4.86 + .17 = 5.03

Subtracting Decimals Mentally

Line up the decimals MENTALLY; then subtract.
Afterwards, write the numbers in the boxes to check your answer.

1. 1.2 - .6 = **.6**

	1	. 2	
-	0	. 6	
		. 6	

CHECK YOUR ANSWER IN THE BOXES AFTER DOING THE PROBLEM MENTALLY!

2. 1.0 - .3 = **.7**

	1	. 0	
-	0	. 3	
		. 7	

3. 20.0 - .7 = **19.3**

	2	0	. 0	
-	0	0	. 7	
	1	9	. 3	

4. 2.57 - 0.3 = **2.27**

	2	. 5	7
-	0	. 3	0
	2	. 2	7

5. 4.25 - 0.3 = **3.95**

	4	. 2	5
-	0	. 3	0
	3	. 9	5

6. 5.4 - 0.25 = **5.15**

	5	. 4	0
-	0	. 2	5
	5	. 1	5

Subtracting Decimals Mentally

Line up the decimals MENTALLY; then subtract.
Afterwards write the numbers in the boxes to check your answer.

1. 4.8 - .7 = **3.1**

2. 1.0 - .6 = **.4**

3. 1.4 - .7 = **.7**

4. 40.0 - .7 = **39.3**

5. 8.57 - 0.3 = **8.27**

6. 3.21 - 0.3 = **2.91**

7. 5.4 - 0.05 = **5.35**

8. 5.4 - 0.25 = **5.15**

9. 3.38 - 0.2 = **6.17**

10. 7.6 - 0.38 = **7.22**

REVIEW: Line up the decimals MENTALLY, then add

11. 2.46 + .3 = **2.76**

12. 2.46 + .02 = **2.48**

13. 2.46 + .32 = **2.78**

14. 5.57 + .06 = **5.63**

15. 3.83 + .80 = **4.63**

16. 3.76 + .75 = **4.51**

Decimal Addition-Subtraction Quiz

Add or subtract the following decimals, following all the steps you have learned so far

1. 7 - 3.62 = **3.38**

```
   7.00
 - 3.62
   3.38
```

2. 7.06 + 35.8 = **42.86**

```
   7.06
 +35.80
  42.86
```

3. 8.5 - 2.07 = **6.43**

```
   8.50
 - 2.07
   6.43
```

4. 95.302 - 23.148 = **72.154**

```
   95.302
 - 23.148
   72.154
```

5. 16.035 - 2.993 = **13.042**

```
   16.035
 - 02.993
   13.042
```

6. 14.008 - 6.037 = **7.971**

```
   14.008
 - 06.037
    7.971
```


Whole Numbers Minus Decimals

Example

A. 4 - .06

Wrong!

```
   4
 - .06
```

THERE IS NO SUCH THING AS "SPACE MINUS 6!"

YOU CAN'T HAVE BLANKS WHEN YOU SUBTRACT!

You have to subtract from zeros instead of blanks

Right!

```
   4.00
 - .06
```

ADD A DECIMAL POINT AND ZEROS ABOVE EVERY DIGIT YOU SUBTRACT FROM!

How to subtract when you have a whole number minus a decimal

#1 Add a decimal point to the right of any whole number

YOU CAN ADD AS MANY ZEROS TO THE ***RIGHT*** OF A DECIMAL AS YOU WANT!

#2 Line up the decimal points and add any placeholder zeros where needed

#3 Subtract normally—and bring down the decimal point

Subtract the following decimals from whole numbers

1. 8 - .4

8	. 0		
0	. 4		
7	. 6		

2. 9 - .76

9	. 0	0	
0	. 7	6	
8	. 2	4	

3. 3 - .185

3	. 0	0	0
0	. 1	8	5
2	. 8	1	5

4. 17 - .26

1	7	. 0	0	
0	0	. 2	6	
1	6	. 7	4	

Concept Quiz 1

1. Tenths are represented by

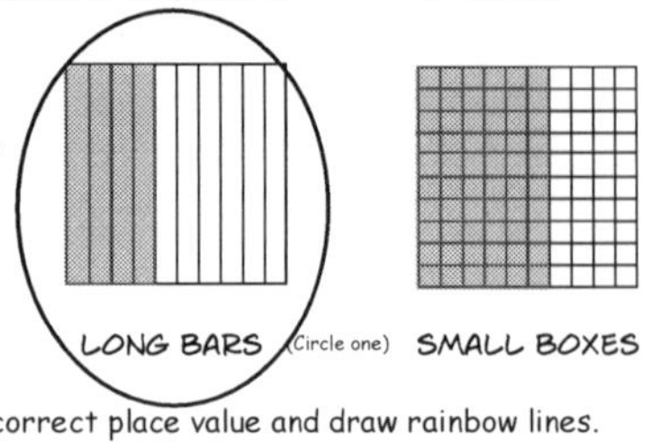

2. Label the boxes with the correct place value and draw rainbow lines.

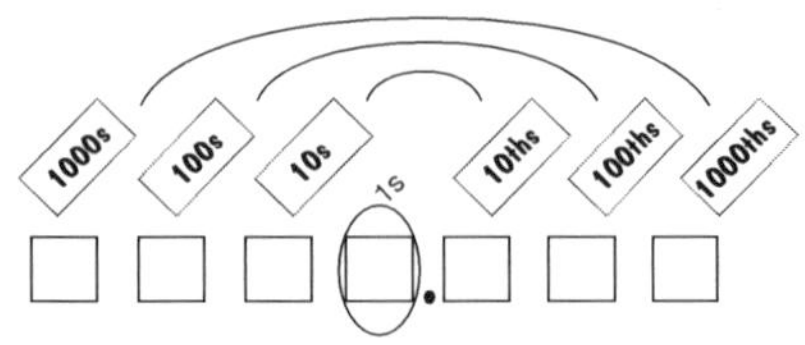

times:	times:	times:		times:	times:	times:
1000	**100**	**10**		**.1**	**.01**	**.001**

3.

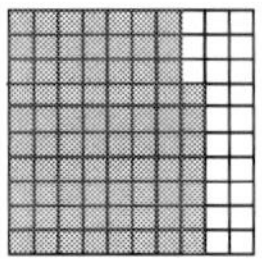

Decimal amount shaded: **.77**

Decimal amount UNshaded: **.23**

WRITE YOUR ANSWERS AS *DECIMALS!*

4. When reading decimal numbers, the decimal point represents this word: **and**

5. Write 24 and 79 thousandths: ☐ ☐ **2** **4** . **0** **7** **9**

Concept Quiz 2

1. What are the three steps you need to take when adding or subtracting decimals (like 7 - .23)?

Step 1: **Add a decimal point to the right of any whole number**

Step 2: **Line up the decimal points and add any placeholder zeros where needed**

Step 3: **Subtract normally—and bring down the decimal point**

2. What is wrong with the following problem? $\begin{array}{r} .75 \\ - \ .3 \\ \hline \end{array}$

The decimal points are not lined up correctly

3. What is missing in the following problem? $\begin{array}{r} 4.2 \\ - \ 2.38 \\ \hline \end{array}$

The 4.2 should be written as 4.20 — you can't have blank minus 8

4. Write the following decimals as fractions: (a) .859 $\boxed{\frac{859}{1000}}$ (b) 2.37 $\boxed{2\frac{37}{100}}$

Show all the steps as you solve these decimal problems:

5. 6.3 - 2.07 = $\begin{array}{r} 6.30 \\ - \ 2.07 \\ \hline 4.23 \end{array}$

6. 7.05 + 5.89 = $\begin{array}{r} 7.05 \\ + \ 5.89 \\ \hline 12.94 \end{array}$

Draw and label points on the number line for the following decimals:

6. .7 **7.** 1.3 **8.** 2.1 **9.** 2.9

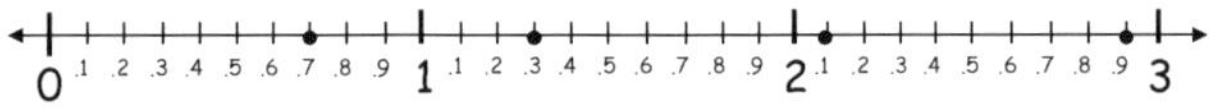

.7 **1.3** **2.1** **2.9**

10. Make a number triangle to show expanded notation . 7 2 8 6

. **7** ☐ ☐ ☐
. **0** **2** ☐ ☐
+ . **0** **0** **8** ☐
+ . **0** **0** **0** **6**

11. Subtract the following decimals 7.1 - 3.83

$\begin{array}{r} 7.10 \\ - \ 3.83 \\ \hline 3.27 \end{array}$

12. The Dizzy Trick - Turn .942 into a fraction

Step one: Write a **zero** under each decimal digit

Step two: Change the decimal point to **1 (to the left)**

Step three: Add a **fraction** line

Answer: $\frac{942}{1000}$

Decimals as Added Fractions

Example

.376 → $\frac{3}{10} + \frac{7}{100} + \frac{6}{1000}$

Write the following decimals as the sum of fractions (as above)

1. .52 — $\frac{5}{10} + \frac{2}{100}$

2. .068 — $\frac{0}{10} + \frac{6}{100} + \frac{8}{1000}$

3. .709 — $\frac{7}{10} + \frac{0}{100} + \frac{9}{1000}$

4. .9837 — $\frac{9}{10} + \frac{8}{100} + \frac{3}{1000} + \frac{7}{10000}$

5. .13 — $\frac{1}{10} + \frac{3}{100}$

6. .003 — $\frac{0}{10} + \frac{0}{100} + \frac{3}{1000}$

7. .176 — $\frac{1}{10} + \frac{7}{100} + \frac{6}{1000}$

8. .042 — $\frac{0}{10} + \frac{4}{100} + \frac{2}{1000}$

Decimals as Added Fractions

Write the following decimals from the sum of fractions

1. .06

$\frac{0}{10} + \frac{6}{100}$

2. .274

$\frac{2}{10} + \frac{7}{100} + \frac{4}{1000}$

3. $.\overline{3}$ (THE LINE ABOVE THE 3 MEANS THAT THE 3 REPEATS TO INFINITY!) = .333...

$\frac{3}{10} + \frac{3}{100} + \frac{3}{1000} + \cdots$

BACKWARDS! NOW YOU WRITE THE DECIMAL!

4. $\frac{4}{10} + \frac{3}{100}$

.4 3

WATCH WHERE YOU PUT THE DECIMAL NUMBERS!

THESE FRACTIONS ARE NOT PUT IN ORDER!

5. $\frac{7}{100} + \frac{2}{10}$

.2 7

6. $\frac{6}{10} + \frac{5}{100} + \frac{2}{1000}$

.6 5 2

7. $\frac{8}{1000} + \frac{1}{100} + \frac{3}{10}$

.3 1 8

8. $\frac{4}{100} + \frac{9}{10} + \frac{7}{1000}$

.9 4 7

9. $\frac{4}{10} + \frac{1}{1000} + \frac{7}{100}$

.4 7 1

10. $\frac{0}{100} + \frac{6}{1000} + \frac{5}{10}$

.5 0 6

Putting Decimals in Order

Example

A. Put the following decimals in order: .1, .009, .08

.1 _ _

.0 0 9 ← 9 is normally larger than 1, but 1 in the tenths place is larger than 9 in the thousandths place!

.0 8 _

↑ Even though 1 is smaller than 9 or 8, it is larger than zero! Placement matters!

Largest decimal: .1

Middle decimal: .08

Smallest decimal: .009

Line up the decimals and put them in order

1. .1 .01 .11

FIRST, COPY THE NUMBERS IN THE ORDER YOU SEE THEM IN THE PROBLEM!

.1 _	Largest decimal:	.11
.0 1	Middle decimal:	.1
.1 1	Smallest decimal:	.01

2. .65 .56 .66

.6 5	Largest decimal:	.66
.5 6	Middle decimal:	.65
.6 6	Smallest decimal:	.56

Putting Decimals in Order

Line up the decimals and put them in order

THIS IS ACTUALLY VERY SIMILAR TO PUTTING WORDS IN ALPHABETICAL ORDER!

Check this column first / Check this column second / Check this column third

YOU CAN LABEL THE DECIMALS "B" FOR BIGGEST, "M" FOR MIDDLE, AND "S" FOR SMALLEST!

1. .09 .2 .089

HERE, JUST PUT THE NUMBERS IN THE ORDER THEY ARE GIVEN!

.0 9 _	Largest decimal:	.2
.2 _ _	Middle decimal:	.09
.0 8 9	Smallest decimal:	.089

2. .07 .777 .7

.0 7 _	Largest decimal:	.777
.7 7 7	Middle decimal:	.7
.7 _ _	Smallest decimal:	.07

3. .181 .180 .118

.1 8 1	Largest decimal:	.181
.1 8 0	Middle decimal:	.180
.1 1 8	Smallest decimal:	.118

4. .207 .270 .072

.2 0 7	Largest decimal:	.270
.2 7 0	Middle decimal:	.207
.0 7 2	Smallest decimal:	.072

Ordering from LEAST to GREATEST

Line up the decimals and put them in order, from LEAST to GREATEST

1. .818 .881 .118

HERE, JUST PUT THE NUMBERS IN THE ORDER THEY ARE GIVEN!

.8 1 8	Smallest decimal:	.118
.8 8 1	Middle decimal:	.818
.1 1 8	Largest decimal:	.881

2. .068 .086 .806

.0 6 8	Smallest decimal:	.068
.0 8 6	Middle decimal:	.086
.8 0 6	Largest decimal:	.806

TIP: LABEL THE DECIMALS 1, 2, 3 (SMALLEST TO LARGEST)!

3. .321 .322 .312

.3 2 1	Smallest decimal:	.312
.3 2 2	Middle decimal:	.321
.3 1 2	Largest decimal:	.322

4. .544 .455 .454

.5 4 4	Smallest decimal:	.454
.4 5 5	Middle decimal:	.455
.4 5 4	Largest decimal:	.544

5. .518 .409 .589

.5 1 8	Smallest decimal:	.409
.4 0 9	Middle decimal:	.518
.5 8 9	Largest decimal:	.589

Concept Quiz

1. Write the number "26 and 318 thousandths"

[ ] [ ] [2] [6] . [3] [1] [8]

2. Put 9 in the hundredths place;
Put 5 in the tenths place;
Put 1 in the thousandths place;
Put 8 in the thousandths place
Put zeros in any remaining places

[8] [0] [0] [0] . [5] [9] [1]

3. 47 ÷ 10 = **4.7**

4. 2.51 · 100 = **251**

5. Write the decimal as a sum of two fractions:

.67

$\frac{6}{10} + \frac{7}{100}$

6. Write the decimal from a sum of fractions:

$\frac{4}{1000} + \frac{5}{100} + \frac{2}{10}$. [2] [5] [4]

7. Line up the decimals and put them and put them in order from least to greatest

.822 .882 .288

. [8] [2] [2] Smallest decimal: **.288**

. [8] [8] [2] Middle decimal: **.822**

. [2] [8] [8] Largest decimal: **.882**

8. 3.5 - 1.02 = **2.48**

9. If you write this decimal as a fraction the bottom number will have a 1 with **3** zeros.

$.856 = \frac{856}{1000}$ ← How many zeros?

10. How do you know how many zeros go on the denominator in #9? (What is the rule?)

One zero for every decimal digit

11. Whenever you DIVIDE by 10 or a power of 10 you slide the decimal to the (LEFT) RIGHT

12. 7.3 - 2.46 = **4.84**

13. 6.37 + .748 = **7.118**

14. Write the following decimal as a sum of fractions:

.976

$\frac{9}{10} + \frac{7}{100} + \frac{6}{1000}$

15. Write the decimal form of this sum of factions

$\frac{2}{10} + \frac{5}{1000} + \frac{8}{100}$

. [2] [8] [5]

16. .001 = $\frac{1}{1000}$ ← [3] zeros

How many zeros will be on the bottom of the fraction?

17. .004 + .2 + .0003 = **.2043**

Multiplying Decimals

Example

TWO SLIDES IN THE PROBLEM... TWO SLIDES IN THE ANSWER!

A. .6 × .3 = .18

COUNT THE TOTAL NUMBER OF SLIDES IN THE PROBLEM... → 2

GIVE YOUR ANSWER THE SAME NUMBER OF SLIDES!

#1 Add up the number of slides in ALL factors

#2 Multiply normally

#3 Give your answer the same number of slides as in #1

YOU CAN JOT DOWN THE NUMBER OF SLIDES IN THE PROBLEM AND THEN ERASE THE DECIMAL POINTS!

...BUT REMEMBER TO GIVE YOUR ANSWER THE SAME NUMBER OF DECIMAL SLIDES!

YOU ONLY LINE THEM UP WHEN YOU ADD OR SUBTRACT THEM!

You don't line up the decimals when you multiply or divide them

1. .3 × .8

#1 Number of slides in the problem: [2]

#2 Multiply normally [3 × 8]

#3 Give your answer the same number of slides: . [2] [4]

.3 × .8 = **.24**

2. .04 × .7

AFTER YOU COUNT THE SLIDES YOU CAN FORGET THEM WHILE YOU CRUNCH THE NUMBERS!

#1 Number of slides in the problem: [3]

#2 Multiply normally [4] × [7] = **28**

#3 Give your answer the same number of slides: . [0] [2] [8]

.04 × .7 = **.028**

DID YOU MAKE ROOM FOR THE EXTRA SLIDE (PUT IN A ZERO)?

Multiplying Decimals

Review

#1 Add up the number of slides in ALL factors

#2 Multiply normally

SERIOUSLY! YOU CAN EVEN PRETEND IT ISN'T A DECIMAL PROBLEM (AT LEAST UNTIL YOU GET TO THE END)!

#3 Give your answer the same number of slides as in the whole problem (same as #1)

Multiply the following decimals by following the steps given

1. .4 × .6

#1 Number of slides in the problem: [2]

#2 Multiply normally [4] × [6] = **24**

#3 Give your answer the same number of slides: .4 × .6 = **.24**

SLIDE LEFT!

2. 7 × .05

MULTIPLY JUST THE WHOLE NUMBERS!

#1 Number of slides in the problem: [2]

#2 Multiply normally [7] × [5] = **35**

#3 Give your answer the same number of slides: 7 × .05 = **.35**

3. .03 × .02

#1 Number of slides in the problem: [4]

#2 Answer when you just multiply the whole numbers: = [6]

#3 Give your answer the same number of slides .03 × .02 = **.0006**

Multiplying Decimals

Multiply the non-zero numbers and give your answer the right number of slides

THE NUMBER OF SLIDES HAS TO MATCH!

1. .3 × 8 = 2.4 — Number of slides: 1
2. .3 × .8 = .24 — Number of slides: 2
3. 7 × .04 = .28 — Number of slides: 2
4. .05 × 7 = .35 — Number of slides: 2
5. .002 × 8 = .016 — Number of slides: 3
6. .04 × 9 = .36 — Number of slides: 2
7. .8 × 4 = 3.2 — Number of slides: 1
8. .6 × .9 = .54 — Number of slides: 2
9. .9 × 5 = 4.5 — Number of slides: 1
10. .7 × .06 = .042 — Number of slides: 3
11. 9 × .07 = .63 — Number of slides: 2

11. .8 × .8 = .64 — Number of slides: 2
12. $.7^2$ = .49 — Number of slides: 2 (.7 • .7)
13. .2 × 10 = 2(.0) — Number of slides: 1
14. .3 × 20 = 6(.0) — Number of slides: 1
15. $.9^2$ = .81 — Number of slides: 2 (.9 • .9)
16. .4 × .30 = .12(0) — Number of slides: 3
17. 1000 × .009 = 9(.0) — Number of slides: 3

Decimal Multiplication Practice

Multiply the non-zero numbers and give your answer the right number of slides

THE NUMBER OF SLIDES HAS TO MATCH!

1. 3 × 5 = 15 — Number of slides: 0
2. .3 × 5 = 1.5 — Number of slides: 1
3. .3 × .5 = .15 — Number of slides: 2
4. .02 × 6 = .12 — Number of slides: 2
5. .01 × 365 = 3.65 — Number of slides: 2
6. 4 × .07 = .28 — Number of slides: 2
7. 6 × .3 = 1.8 — Number of slides: 1
8. .2 × .2 × .2 = .008 — Number of slides: 3
9. .1 × 10 = 1 — Number of slides: 1
10. .01 × 100 = 1 — Number of slides: 2
11. .3 × .02 = .006 — Number of slides: 3

12. .3 × .3 = .09 — Number of slides: 2
13. $.5^2$ = .25 — Number of slides: 2

HINT! -- WRITE THIS OUT AS MULTIPLICATION PROBLEM!

14. .06 × 100 = 6 — Number of slides: 2
15. .8 × 30 = 24 — Number of slides: 1
16. $.2^2$ = .04 — Number of slides: 2
17. .1 × .1 × .1 × .1 = .0001 — Number of slides: 4
18. .03 × .04 = .0012 — Number of slides: 4

Decimal Multiplication Practice

Multiply the decimals and give your answer the right number of slides

THE NUMBER OF SLIDES HAS TO MATCH!

WHEN YOU MULTIPLY BY *LESS THAN ONE*, YOU ARE REALLY *DIVIDING!*

1. .2 × 4 = .8
2. .2 × .4 = .08
3. .03 × .2 = .006
4. .001 × 7 = .007
5. 5 × .3 = 1.5
6. .1 × .1 = .01
7. 8 × .07 = .56
8. .2 × .9 = .18
9. .06 × .1 = .006
10. .3 × .09 = .027
11. .7 × 4 = 2.8
12. .6 × .02 = .012
13. .09 × .02 = .0018
14. .3 × .12 = .036
15. .05 × .2 = .01(0)
16. .7 × .07 = .049
17. 6 × .7 = 4.2
18. .08 × .02 = .0016
19. 3 × .09 = .27
20. .2 × .2 × .3 = .012
21. .2 × .3 × .04 = .0024
22. .3 × .3 × 6 = .54

Decimal Multiplication Practice

Multiply the decimals and give your answer the right number of slides

When you multiply by LESS than ONE, you are really ____________

1. .3 × .7 = .21
2. .3 × 7 = 2.1
3. .8 × .3 = .24
4. .12 × .3 = .036
5. .04 × .2 = .008
6. .9 × .09 = .081
7. 5 × .006 = .03(0)
8. 6 × .8 = 4.8
9. .08 × .02 = .0016
10. .01 × .039 = .00039
11. .4 × .006 = .0024
12. .2 × .3 × 10 = .6

MULTIPLYING BY 10 MOVES THE DECIMAL ONE TIME TO THE RIGHT!

13. .3 × 6 ÷ 10 = .18

DIVIDING BY 10 MOVES THE DECIMAL ONE TIME TO THE LEFT!

14. .7 × .3 ÷ 10 = .021
15. .8 × .6 × 10 = 4.8
16. .4 × .4 ÷ 10 = .016
17. .5 × 9 × 10 = 45
18. .3 × .9 ÷ 10 = .027
19. .5 × .07 × 10 = .35
20. .08 × .02 × 100 = .16

Multiplying Decimals

Examples

A.

73
× .3 ← 1 slide
29.2 ← 1 slide

ONE SLIDE IN THE PROBLEM...

ONE SLIDE IN THE ANSWER!

B.

.36 ← 2 slides
× .27 ← 2 slides
252
820
.1072 ← 4 slides

IGNORE THE DECIMALS AS YOU WORK THE PROBLEM!

JUST REMEMBER TO GIVE YOUR ANSWER THE SAME NUMBER OF SLIDES AS IN THE PROBLEM!

#1 Count the number of slides in the problem

#2 Multiply normally

#3 Give your answer the same number of slides

1. 4.6 ← 1 slide
× 3.2 ← 1 slide
1472 Multiply normally (ignore the decimals)
14.72 Give your answer 2 slides

PUT YOUR ANSWER (WITH DECIMAL SLIDES) HERE!

2. .64 ← 2 slides
× 2.5 ← 1 slide
1600 Multiply normally (ignore the decimals)
1.600 or 1.6 Give your answer **3** slides

3. 74
× .02 ← **2** slide(s)
148 Multiply normally (ignore the decimals)
1.48 Give your answer **2** slides

4. .07 ← **2** slide(s)
× .3 ← **1** slide(s)
21 Multiply normally (ignore the decimals)
.021 Give your answer **3** slides

5. 284 ← **0** slide(s)
× .04 ← **2** slide(s)
1136 Multiply normally (ignore the decimals)
11.36 Give your answer **2** slides

6. 1.76 ← **2** slide(s)
× .23 ← **2** slide(s)
4048 Multiply normally (ignore the decimals)
.4048 Give your answer **4** slides

Multiplying Decimals

1. .57 ← **2** slide(s)
× 3.5 ← **1** slide(s)
1995 Multiply normally (ignore the decimals)
1.995 Give your answer **3** slides

PUT YOUR ANSWER (WITH DECIMAL SLIDES) HERE!

2. .36 ← **2** slide(s)
× .8 ← **1** slide(s)
288 Multiply normally (ignore the decimals)
.288 Give your answer **3** slides

3. 7.58 ← **2** slide(s)
× .26 ← **2** slide(s)
19708
19.708 Give your answer **4** slides

4. 83 ← **0** slide(s)
× .03 ← **2** slide(s)
249
2.49 Give your answer **2** slides

5. 2.3 ← **1** slide(s)
× 6.1 ← **1** slide(s)
1403
14.03 Give your answer **2** slides

6. .48 ← **2** slide(s)
× 6.7 ← **1** slide(s)
3216
3.216 Give your answer **3** slides

7. 631 ← **0** slide(s)
× .24 ← **2** slide(s)
15,144
151.44 Give your answer **2** slides

8. .59 ← **2** slide(s)
× .43 ← **2** slide(s)
2537
.2537 Give your answer **4** slides

Multiplying Decimals

1. 14.6 ← **1** slide(s)
× 4.2 ← **1** slide(s)
6132
61.32 Give your answer **2** slides

2. 1.62 ← **2** slide(s)
× 7.5 ← **1** slide(s)
12,150
12.15(0) Give your answer **3** slides

3. 4.2 ← **2** slide(s)
× .09 ← **1** slide(s)
378
.378 Give your answer **3** slides

4. 8.5 ← **1** slide(s)
× 7.6 ← **1** slide(s)
6460
64.60 Give your answer **2** slides

5. 2.3 ← **1** slide(s)
× 6.4 ← **1** slide(s)
1472
14.72 Give your answer **2** slides

6. 13.2 ← **1** slide(s)
× 9 ← **0** slide(s)
1188
118.8 Give your answer **1** slides

7. 10.2 ← **1** slide(s)
× .67 ← **2** slide(s)
6834
6.834 Give your answer **3** slides

8. 5.8 ← **1** slide(s)
× 3.3 ← **1** slide(s)
1914
19.14 Give your answer **2** slides

Concept Quiz

1. Whenever you have a problem like [7 - .03] you have to make one slight change before you can subtract the numbers. What is that change?

Add a decimal point and two zeros to the 7 (7.00)

(This is more important with subtraction than with addition.)

2. Why do you have to do this when you have this type of decimal subtraction problem?

You can't have blank minus 3, etc.

3. Solve: [6 - .07] Answer: [**5.93**]

4. Whenever you have a decimal multiplication problem like .3 x .7 you have to do these three steps:

Step 1: **Count the decimal slides from ALL the factors**

Step 2: **Take off the decimals and multiply normally**

Step 3: **Give your answer the same number of decimal slides as in Step 1**

5. .07 is the same as 7 DIVIDED by **100**

6. .07 is the same as 7 MULTIPLIED by $\frac{1}{100}$

7. When you MULTIPLY decimals you DON'T line up the decimal points first.

(TRUE) FALSE

8. To go from a decimal to a percent or a percent to a decimal you have to slide **2** times.

9. What in the percent sign tells you this?

The two zeros

10. If you have a PERCENT and you want a DECIMAL you slide the decimal to the

RIGHT (LEFT)

Percent - Decimal Conversions

11. 7% = **.07** (decimal)

12. .06 = **6** % (percent)

13. .003 = **.3** % (percent)

14. .25 = **25** % (percent)

DM-10 Problems

15. 54 ÷ 10 = **5.4**

16. .932 × 100 = **93.2**

Decimal Multiplication

17. .6 × .08 = **.048**

18. 2.49 × 5.3 = **13.197**

```
   2.4 9
 ×   5.3
   7 4 7
 1 2 4 5 0
 1 3 1 9 7
```


Decimal Dividends

Example

A.
```
    4.2
2 ) 8↑4
```

KEEP THE QUOTIENT ANSWERS ABOVE THE RIGHT NUMBERS!

...CALLED THE *DIVIDEND!*

IT FLOATS UP TO THE TOP!

Move the decimal straight up

Steps:

#1 Divide normally. (Be sure to line up the quotient answers correctly.)

#2 After you have divided the numbers, just give your answer a decimal point right above the decimal in the dividend.

Solve the following decimal division problems

1. 2) 82↑6 — **41.3**

2. 3) 9↑63 — **3.21**

3. 2) 7↑34 — **3.67**

4. 4) 64↑8 — **16.2**

5. 4) 64.8 — **16.2**

6. 2) 1.046 — **0.523**

Decimal Dividends

Solve the following decimal division problems

These kinds of problems are common when converting FRACTIONS to DECIMALS

1. 2) 1↑0 — **0.5**

2. 5) 3↑0 — **0.6**

3. 4) 1↑00 — **0.25**

4. 8) 1.000 — **0.125**

5. 4) 3↑00 — **0.75**

6. 3) 1↑000 — **0.333 $\frac{1}{3}$**

Just solve this to three places

7. 3) 2↑000 — **0.666 $\frac{2}{3}$**

Just solve this to three places

8. 8) 7.000 — **0.875**

Decimal Divisors

Examples

A. 1.2) 2.4 = 12) 24 — 2

Get rid of the decimal in the divisor by sliding the decimal to the right of the 1s place

Slide here too

B. .3) 36 = 3) 360. — 3

You may need to add zeros to slide the decimal

(remember the invisible decimal point to the right of the one's place)

DIVISION PROBLEMS ARE JUST LIKE FRACTIONS—YOU CAN MULTIPLY BOTH SIDES BY ANY NUMBER!

In these examples you are just multiplying both sides by 10

Rewrite and solve the following decimal division problems

1. 1.1) 3.3 → **11) 33** — **3**

the divisor needs one slide

$\frac{3.3}{1.1}$ (× 10) = $\frac{33}{11}$

SINCE YOU'RE MULTIPLYING BOTH TOP AND BOTTOM BY THE SAME NUMBER, YOU AREN'T CHANGING THE ANSWER AT ALL!

2. .12) 48 → **12) 4800** — **400**

the divisor needs two slides

3. .6) 30 → **6) 300** — **50**

4. .007) 42 → **7) 42000** — **6000**

5. .2) 8 → **2) 80** — **40**

6. .001) 6 → **1) 6000** — **6000**

7. .05) 35 → **5) 3500** — **700**

8. .03) 60 → **3) 6000** — **2000**

Decimal Divisors

Rewrite and solve the following decimal division problems

THE AMOUNT YOU SLIDE DEPENDS ON THIS NUMBER!

Slide both numbers and divide

1. $1.2\overline{)48}$ — the divisor needs one slide — $12\overline{)480}$ = **40**
2. $.08\overline{)72}$ — the divisor needs two slides — $8\overline{)7200}$ = **900**
3. $.7\overline{).49}$ — $7\overline{)4.9}$ = **.7**
4. $.4\overline{).452}$ — $4\overline{)4.52}$ = **1.13**
5. $.006\overline{).0078}$ — $6\overline{)7.8}$ = **1.3**
6. $.011\overline{)88.44}$ — $11\overline{)88440}$ = **8040**
7. $.05\overline{)846.5}$ — $5\overline{)84650}$ = **16930**
8. $.008\overline{)2.72}$ — $8\overline{)2720}$ = **340**

Why Multiplying Decimal Divisors Works

When you slide decimal divisors it's just like multiplying fractions by the same number

Example

A. $1.2\overline{)48}$

BOTTOM NUMBER OF A FRACTION (1.2)

TOP NUMBER OF A FRACTION (48)

The TOP number is the one that wears the "division hat"

The outside number ("divisor") goes on the BOTTOM of the fraction

$\frac{48}{1.2}$ (× 10) = $\frac{480}{12}$

NOW WE NO LONGER HAVE A DECIMAL IN THE FRACTION (DIVISION PROBLEM)!

Rewrite and the following decimal division problems as fractions and get rid of the decimal points by multiplying

	Convert to a fraction	Multiply by 10, 100, 1000, etc. to get rid of the decimal point		Fraction without a decimal point
1. $.5\overline{)3.5}$ (the divisor needs one slide)	**3.5** / **.5**	× **10** / × **10**	=	**35** / **5**
2. $.09\overline{)27}$ (the divisor needs two slides)	**27** / **.09**	× **100** / × **100**	=	**2700** / **9**
3. $.006\overline{)4.2}$ (the divisor needs three slides)	**4.2** / **.006**	× **1000** / × **1000**	=	**4200** / **6**

.5 is the Same as One Half

.5 is the same as 1/2 (divide by 2)

Example

A. .5 × 6 = 3

.5 IS ONE HALF; WHAT'S HALF OF 6?

.5 IS THE SAME AS 50%!

Multiply the following numbers by .5

1. .5 × 8 = **4**
2. .5 × 16 = **8**
3. .5 × 10 = **5**
4. .5 × 20 = **10**
5. .5 × 18 = **9**
6. .5 × 22 = **11**
7. .5 × 100 = **50**
8. .5 × 220 = **110**
9. .5 × 60 = **30**
10. .5 × 480 = **240**

DECIMAL REVIEW

11. Which is larger: .7 or .08? **.7** Why? **.7 in the 10ths place is greater than 0 in the tenths place**
12. Write the number "three and 42 thousandths" **3.042**

.25 is the Same as One Fourth

.25 is the same as 1/4 (divide by 4)

Example

A. .25 × 40 = 10

.25 IS ONE FOURTH; WHAT'S A FOURTH OF 40?

.25 IS THE SAME AS 25%!

Multiply the following numbers by .25 or .5

1. .25 × 8 = **2**
2. .25 × 20 = **5**
3. .25 × 12 = **3**
4. .25 × 24 = **6**
5. .25 × 40 = **10**
6. .5 × 20 = **10**
7. .25 × 100 = **25**
8. .25 × 88 = **22**
9. .5 × 16 = **8**
10. .25 × 16 = **4**
11. .5 × 24 = **12**
12. .25 × 24 = **6**
13. .25 × 28 = **7**
14. .25 × 20 = **5**
15. .25 × 64 = **16**
16. .25 × 144 = **36**

Decimal Multiplication Review

1. 7.2 ← **1** slide(s)
× 5.9 ← **1** slide(s)
4248
42.48 Give your answer **2** slides

5. 3.8 3 ← **2** slide(s)
× 6.4 ← **1** slide(s)
24,512
24.512 Give your answer **3** slides

2. 48.1 ← **1** slide(s)
× 6 ← **0** slide(s)
2886
288.6 Give your answer **1** slides

6. 3.7 ← **1** slide(s)
× .0 9 ← **2** slide(s)
333
.333 Give your answer **3** slides

3. 57.9 ← **1** slide(s)
× . 4 3 ← **2** slide(s)
24,897
24.897 Give your answer **3** slides

7. 8.0 4 ← **2** slide(s)
× . 2 4 ← **2** slide(s)
19,296
1.9296 Give your answer **4** slides

4. .9 6 ← **2** slide(s)
× 8.2 ← **1** slide(s)
7872
7.872 Give your answer **3** slides

8. 4.8 ← **1** slide(s)
× 7.5 ← **1** slide(s)
3600
36.(00) Give your answer **2** slides

Decimals Test 1

1. Whenever you have a problem like $.2\overline{)6}$ you first have to
Eliminate the decimal in the divisor (.2 → 2)

2. Write $.2\overline{)6}$ as a fraction **$\frac{6}{.2}$**

3. Solve: $.07\overline{)63}$ Answer: **900**

$7\overline{)6300}$

4. When you multiply by .5 it's the same as **Times 1/2 (or divide by 2)**

5. .5 × 48 = **24**

6. When you multiply by .25 it's the same as **Times 1/4 (or divide by 4)**

7. .25 × 32 = **8**

Decimals Test 2

1. Whenever you have a problem like 2.4 + .7 8 you first have to
Line up the decimals

2. Solve: 3.7 5 + 4.7 = **8.45**

3. Whenever you have a problem like $.2\overline{)6}$ you first have to
Multiply both .2 and 6 by 10 (to eliminate the decimal divisor)

4. **0.8 7 9**
$5\overline{)4.395}$

5. Solve: 4.2
× .3 7
1.554

6. .734 is the same as 734 DIVIDED by **1000**

. . . or 734 MULTIPLIED by $\frac{1}{\mathbf{1000}}$

8. Whenever you multiply decimals you have to line up the decimals first

TRUE (FALSE)

Now that you've learned the basics of decimals, look for

MathWise Percents

or

MathWise Fractions

...with more tips & tricks!

Made in United States
Troutdale, OR
11/03/2026

56844445R00060